## NASA STI Program…in Profile

Since its founding, NASA has been dedicated to the advancement of aeronautics and space science. The NASA scientific and technical information (STI) program plays a key part in helping NASA maintain this important role.

The NASA STI program operates under the auspices of the Agency Chief Information Officer. It collects, organizes, provides for archiving, and disseminates NASA's STI. The NASA STI program provides access to the NASA Aeronautics and Space Database and its public interface, the NASA Technical Report Server, thus providing one of the largest collections of aeronautical and space science STI in the world. Results are published in both non-NASA channels and by NASA in the NASA STI Report Series, which includes the following report types:

- **Technical Publication:** Reports of completed research or a major significant phase of research that present the results of NASA Programs and include extensive data or theoretical analysis. Includes compilations of significant scientific and technical data and information deemed to be of continuing reference value. NASA counterpart of peer-reviewed formal professional papers but has less stringent limitations on manuscript length and extent of graphic presentations.
- **Technical Memorandum:** Scientific and technical findings that are preliminary or of specialized interest, e.g., quick release reports, working papers, and bibliographies that contain minimal annotation. Does not contain extensive analysis.
- **Contractor Report:** Scientific and technical findings by NASA-sponsored contractors and grantees.
- **Conference Publication:** Collected papers from scientific and technical conferences, symposia, seminars, or other meetings sponsored or co-sponsored by NASA.
- **Special Publication:** Scientific, technical, or historical information from NASA programs, projects, and missions, often concerned with subjects having substantial public interest.
- **Technical Translation:** English-language translations of foreign scientific and technical material pertinent to NASA's mission.

Specialized services also include creating custom thesauri, building customized databases, and organizing and publishing research results.

For more information about the NASA STI program, see the following:

- Access the NASA STI program home page at *http://www.sti.nasa.gov*
- E-mail your question via the Internet to *help@sti.nasa.gov*
- Fax your question to the NASA STI Help Desk at 443-757-5803
- Phone the NASA STI Help Desk at 443-757-5802

Write to:

NASA STI Help Desk
NASA Center for AeroSpace Information
7115 Standard Drive
Hanover, MD 21076-1320

NASA/SP-2011-3423

# NASA ACCIDENT PRECURSOR ANALYSIS HANDBOOK

## Version 1.0

National Aeronautics and Space Administration
Office of Safety and Mission Assurance
Washington, D.C. 20546

December 2011

# Table of Contents

Acknowledgments ..................................................................................................................... v
1 Introduction ........................................................................................................................ 1
   1.1 Background ................................................................................................................ 1
   1.2 Summary of Accident Precursor Analysis ................................................................ 1
   1.3 History of NASA's Precursor Program ..................................................................... 2
   1.4 Handbook Overview ................................................................................................. 3
2 Accident Precursor Analysis Overview .............................................................................. 5
   2.1 The Accident Precursor Concept ............................................................................... 5
   2.2 Accident Precursor Analysis and its Role in System Safety .................................... 8
   2.3 NASA Accident Precursor Analysis Process Overview ......................................... 12
3 Accident Precursor Analysis Process Steps ..................................................................... 19
   3.1 Building a Caseload ................................................................................................ 19
      3.1.1 Data Sources of Interest ................................................................................ 19
      3.1.2 Choosing Accident Precursor Analysis Relevant Data ................................ 19
      3.1.3 Timetable for Evaluation of Anomaly Data ................................................. 21
      3.1.4 Anomaly Screening Methods ....................................................................... 22
   3.2 Anomaly Failure Mechanism Identification & Generalization .............................. 23
      3.2.1 Identifying the Anomaly Failure Mechanism .............................................. 24
      3.2.2 The Scope of Generalization ........................................................................ 25
      3.2.3 The Generalization Group ............................................................................ 26
      3.2.4 Anatomy of the Anomalous Condition Accident Sequence ........................ 28
      3.2.5 Generalizing the Anomaly Failure Mechanism ........................................... 29
      3.2.6 The Generalization Team ............................................................................. 32
   3.3 Evidence Gathering ................................................................................................. 34
   3.4 Grading of Anomalous Conditions ......................................................................... 37
      3.4.1 Assigning the Failure Condition Index ........................................................ 39
      3.4.2 Assigning the Conditional Consequence Index ........................................... 43
      3.4.3 Grading Results ............................................................................................ 44
      3.4.4 The Grading Team ....................................................................................... 47
   3.5 Observation & Trending ......................................................................................... 47
      3.5.1 Using Trending Data as Evidence ................................................................ 49
      3.5.2 Trending Parameters .................................................................................... 50
   3.6 Risk Modeling ......................................................................................................... 55
      3.6.1 Anomalous Condition Risk Importance ...................................................... 55
      3.6.2 Risk Modeling Outcomes ............................................................................. 57
4 Documentation of Process Steps and Reporting Results ................................................. 58
   4.1 Reports from DAnGERS .......................................................................................... 59
5 Application to Other Mission Classes .............................................................................. 60
6 Conclusion ........................................................................................................................ 62

Appendix A: Acronyms ........................................................................................................... 65
Appendix B: References .......................................................................................................... 67

Appendix C: Tool for Deliberative Anomaly Grading and Evaluation ............................ 69
Appendix D: DanGERS Reports ............................................................................... 88
Appendix E: Defining a Trending Basis ..................................................................... 95
Appendix F: Technical Basis for the Anomalous Condition Risk Index (ACRI) ........... 99
Appendix G: Sample APA Results Reporting .......................................................... 103

# Table of Figures

Figure 2-1 The Swiss Cheese Model of Accident Causation ........................................... 5
Figure 2-2 Post-flight Inspection Photos of TPS Damage on STS-45 ............................. 7
Figure 2-3 Real World vs. Models Information Flow in the APA Context ..................... 10
Figure 2-4 System Safety Claim Model .......................................................................... 11
Figure 2-5 NASA Accident Precursor Analysis Process Overview Diagram ................. 13
Figure 2-6 Timeline of APA Process Steps .................................................................... 14
Figure 3-1 Anomaly Generalization Illustration ............................................................. 24
Figure 3-2 Many-to-many Relationship between Anomalies and Anomalous Conditions ............................................................................................................................................ 27
Figure 3-3 Recurring Anomalies in Generalization Groups ........................................... 27
Figure 3-4 Structure of an Anomalous Condition Accident Sequence ........................... 28
Figure 3-5 Anomalous-Condition-Dependent Consequence Potential ........................... 32
Figure 3-6 Example Summary of Evidence Gathered .................................................... 37
Figure 3-7 FCI and CCI in the accident sequence .......................................................... 38
Figure 3-8 Notional Graph of Recommended Further Action as a Function on FCI, CCI, and $EC_{PPI}$ ....................................................................................................................... 46
Figure 3-9 Generalization Groups and Trending Analysis ............................................. 48
Figure 3-10 Trending Informs the Group ....................................................................... 49
Figure 3-11 Trending Data Feedback Process ................................................................ 50
Figure 3-12 Sample Anomaly Fault Magnitude Trends ................................................. 52
Figure 3-13 PPI Trend of a Generalization Group .......................................................... 53
Figure 3-14 Sample Duane & Cumulative Frequency Plots ........................................... 54

# Table of Tables

Table 2.1 NASA APA Process Steps and Required Skill Sets ....................................... 15
Table 3.1 Common Data Types and Minimum Criteria for Inclusion of Data ............... 36
Table 3.2 FCI Assignment Table ..................................................................................... 40
Table 3.3 CCI Assignment Table ..................................................................................... 43

# Acknowledgments

Project Manager:     Frank Groen, PhD
                     NASA Office of Safety and Mission Assurance (OSMA)

The NASA Accident Precursor Analysis handbook was co-authored by Chris Everett, Tony Hall, and Scott Insley of Information Systems Laboratories (ISL).

Dr. Michael Stamatelatos and Dr. Homayoon Dezfuli (OSMA) initiated the development of the precursor methodology for NASA applications and provided valuable technical inputs and feedback throughout its development. At the formative stages of the effort, important conceptual contributions were made by Bill Vesely (OSMA), Bob Youngblood and Gaspare Maggio, both previously with ISL, and Joe Fragola (Valador).

The development of the methodology benefitted greatly from working sessions with the Space Shuttle and International Space Station Programs that were organized with support from Mike Canga and Mike Lutomski of those respective programs.

U.S. Nuclear Regulatory Commission staff, specifically Pat Baranowsky, Gary DeMoss, and Don Marksberry, supported the development of this handbook by sharing lessons learned from USNRC's application of precursor analysis to nuclear plant safety and observing pilot application working sessions.

Others that contributed to this effort include:

- The JSC team led by Roger Boyer and Teri Hamlin;
- The MSFC team led by Rob Ring (BTI);
- The ISS precursor team led by Alicia Carrier (SAIC);
- The ARC Mission Assurance Systems team including Irene Tollinger, Christian Ratterman, Don Kalar, and Alex Eiser;
- Clay Smith (APL)

# 1 Introduction

## 1.1 Background

Catastrophic accidents are usually preceded by precursory events that, although observable, are not recognized as harbingers of a tragedy until after the fact. In the nuclear industry, the Three Mile Island accident was preceded by at least two events portending the potential for severe consequences from an underappreciated causal mechanism [1]. Anomalies whose failure mechanisms were integral to the losses of Space Transportation Systems (STS) *Challenger* and *Columbia* had been occurring within the STS fleet prior to those accidents. Both the Rogers Commission Report [2] and the *Columbia* Accident Investigation Board report [3] found that processes in place at the time did not respond to the prior anomalies in a way that shed light on their true risk implications.

This includes the concern that, in the words of the NASA Aerospace Safety Advisory Panel (ASAP) [4], "no process addresses the need to update a hazard analysis when anomalies occur." At a broader level, the ASAP noted in 2007 [5] that NASA "could better gauge the likelihood of losses by developing leading indicators, rather than continue to depend on lagging indicators".

These observations suggest a need to revalidate prior assumptions and conclusions of existing safety (and reliability) analyses, as well as to consider the potential for previously unrecognized accident scenarios, when unexpected or otherwise undesired behaviors of the system are observed. This need is also discussed in NASA's system safety handbook [6], which advocates a view of safety assurance as driving a program to take steps that are necessary to establish and maintain a valid and credible argument for the safety of its missions.

It is the premise of this handbook that making cases for safety more experience-based allows NASA to be better informed about the safety performance of its systems, and will ultimately help it to manage safety in a more effective manner.

## 1.2 Summary of Accident Precursor Analysis

The APA process described in this handbook provides a systematic means of analyzing candidate accident precursors by evaluating anomaly occurrences for their system safety implications and, through both analytical and deliberative methods used to project to other circumstances, identifying those that portend more serious consequences to come if effective corrective action is not taken. APA builds upon existing safety analysis processes currently in practice within NASA, leveraging their results to provide an improved understanding of overall system risk. As such, APA represents an important dimension of safety evaluation; as operational experience is acquired, precursor information is generated such that it can be fed back into system safety analyses to risk-inform safety improvements. Importantly, APA utilizes anomaly data to predict risk whereas standard reliability and PRA approaches utilize failure data which often is limited and rare.

The purpose of the APA process is to identify and characterize potential sources of safety risk for which indications are received in the form of anomalous events which, although not necessarily presenting an immediate safety impact, may indicate that an unknown or insufficiently understood potential risk-significant condition exists in the system. Such anomalous events are considered to be potential accident precursors because they signal the potential for more severe consequences that may occur in the future, due to failure mechanisms that are discernible from their occurrence today. Their early identification allows them to be fully scrutinized and the results to be used to inform decisions relating to safety. Stemming from the anomalous event that was actually observed, the NASA process invokes an "imaginative" aspect to the process using a structured brainstorming session to identify similar anomalous conditions which could have more severe consequences than the observed anomalous event. In the context of NASA systems, the term *severe consequences* typically refers to loss of crew (LOC), loss of vehicle (LOV), loss of mission (LOM), or loss of science (LOS). It is up to the particular program employing the approach to define severe consequences appropriate to its objectives and apply the technical approach accordingly.

The APA process presented in this document has been applied to earth-to-orbit transportation systems and crewed orbital science platforms, although the fundamental process steps are valid for other mission classes (e.g., crewed and uncrewed orbital platforms, crewed lunar and planetary outposts, deep-space robotic missions, and other human space exploration missions), and may be tailored to the specific needs of each class. Programs at NASA that have benefited from the APA process presented in this document include the Space Shuttle and the ISS. In addition, NASA is continuing to exercise a robust terrestrial and solar system satellite and robotic based science agenda that could benefit from a systematic APA process. In this case, an accident precursor process could provide valuable information to guide the design of future scientific missions as well as indicate when corrective actions are required during the mission to preclude potential mission-ending failures. Finally, APA plays an important role in extending NASA's anomaly management process to provide additional screening and assessment of anomalies for their risk significance.

## 1.3 History of NASA's Precursor Program

In February 2007 the NASA Office of Safety and Mission Assurance (OSMA) hosted a "Precursor Analysis Working Group Kick-off Meeting" to discuss the development of an Accident Precursor Analysis (APA) process at NASA. Shortly after, an APA team was formed with the intention of utilizing the U.S. Nuclear Regulatory Commission's (NRC) Accident Sequence Precursor (ASP) process [7] as a point of departure for the development of a NASA-specific process, augmented as necessary based on fundamental differences in the nature of the two organizations. In particular, the process presented in this document makes use of NASA's data-rich environment and is tailored to the high-performance space systems that the agency designs and operates.

A first version of an APA approach tailored to NASA's needs, derived from the NRC ASP process and contributions by Dr. Bill Vesely [8, 9], was completed in 2008 [10].

The approach was tested and refined based on a number of preliminary and on-site pilot exercises. First, using an early draft of the process, a retrospective APA assessment was conducted on the significant Thermal Protection System (TPS) damage and the major External Tank (ET) foam loss incidents that occurred prior to Columbia that were identified by the Columbia Accident Investigation Board [3]. Second, a number of APA working sessions were conducted at the Johnson Space Center (JSC) to serve as pilot applications, in collaboration with the Space Shuttle and International Space Station programs [11]. Following those pilot exercises, both programs independently conducted precursor exercises.

This handbook captures the experiences and lessons learned from the above activities.

## *1.4 Handbook Overview*

**Section 2 - Accident Precursor Analysis Overview**, presents a summary background and overview of the NASA APA process. This section outlines the sequence of steps involved in screening, generalization, grading, risk modeling, and reporting of findings. It presents the technical and risk management rationale behind the approach, and the benefit that APA brings to risk management.

**Section 3 -Accident Precursor Analysis Process Steps**, details the sequence of tasks required to conduct a full APA cycle. Divided into the following sub-sections;

> **3.1 - Building a Caseload**, addresses the collection of anomaly source data. This section touches on the use of existing problem reporting data sources, the use of multiple data sources, and the timing of caseload assembly with respect to the initial reporting of the anomalies and subsequent investigatory activities. It also addresses the use of screening methods to filter out anomalies with little or no potential for more severe consequences.

> **3.2 - Anomaly Failure Mechanism Identification & Generalization**, discusses the process of extrapolating an anomaly's underlying failure mechanism and applying it to other circumstances under which it might recur, such as on a different sub-system, at a different time, or with a different fault magnitude. It also addresses the makeup and conduct of a generalization session.

> **3.3 - Evidence Gathering**, addresses the collection of relevant data to support subsequent grading of anomalous conditions, such as typical types of supporting information and criteria by which data can be considered relevant to APA.

> **3.4 - Grading of Anomalous Conditions**, addresses the activity of grading each anomalous condition as requiring either risk modeling, observation and trending, or no further analysis. It also explains the role that evidence plays in determining the pathway to which anomalous conditions are graded.

**3.5 -Observation & Trending**, addresses details the activity of identifying suitable populations for trending analysis, trending anomaly parameters that may signify deteriorating safety levels, and integrating trending results into the APA process.

**3.6 - Risk Modeling**, discusses how scenario-based risk models can be used to assess risk significance, and to identify vulnerabilities in the modeled system. It introduces a number of importance measures that can be used to generate focused recommendations for further analysis and/or testing.

**Section 4 - Documentation of Process Steps and Reporting Results**, summarizes the documentation defined as part of the APA process and various methods of effectively communicating results.

Finally, the main body of the Handbook ends with **Section 5, Conclusion**.

Throughout the document, yellow APA Example boxes illustrate the concepts presented in the various subsections in which they arise, in order to communicate a more concrete understanding of the material and its practical application. The blue boxes peppered throughout the handbook define important APA terms and operations.

# 2 Accident Precursor Analysis Overview

## 2.1 The Accident Precursor Concept

The Swiss Cheese Model of accident causation, originally proposed by James Reason [12], likens a system's barriers against severe failure to a series of slices of randomly-holed Swiss cheese arranged parallel to each other. Each slice could represent a safety process, preventative maintenance, a functional redundancy, etc. The holes represent latent conditions, possible severe stresses, opportunities for human error, adverse environmental conditions, or simply specific subsystem failures. Essentially, the holes in the cheese slices represent inherent vulnerabilities in the system to various events and conditions, and are continually varying in size and position in all slices. Using the Swiss Cheese Model, an accident can be represented as a trajectory through a momentary alignment in a set of holes (as shown by the red line in Figure 2-1). In other words, the causal failure mechanism can sequentially negotiate these holes thus compromising a barrier meant to obviate catastrophe and snowball to a full-blown accident. Whenever a failure mechanism manages to make it through one or more holes, but not all, it is effectively deflected from continuing to a severe consequence (as shown by the blue line in Figure 2-1) and it is cataloged as an *anomaly*.

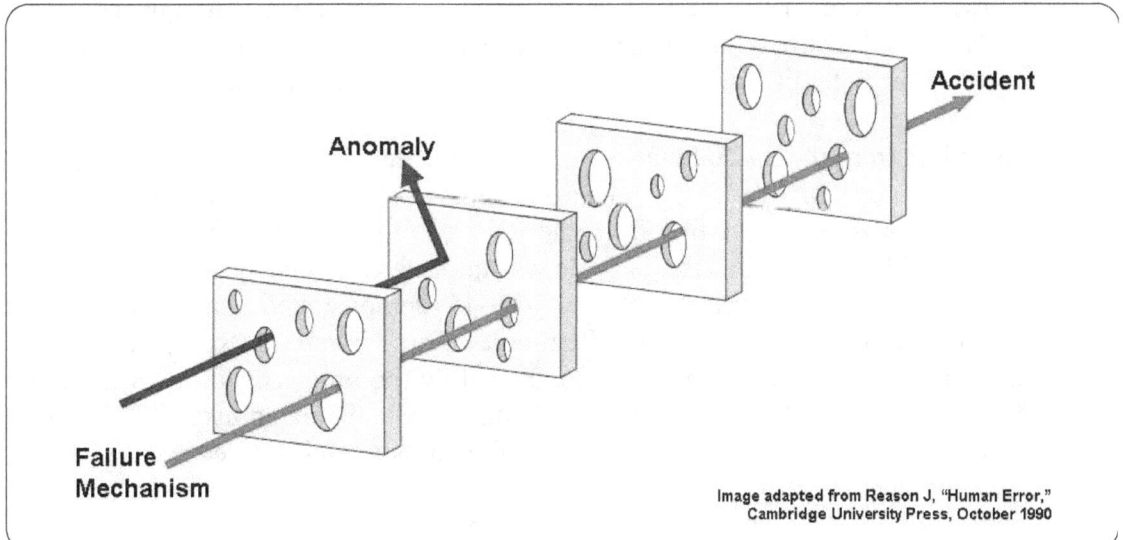

**Figure 2-1 The Swiss Cheese Model of Accident Causation**

An anomaly can make an organization aware of failure mechanisms in the system that may, in combination with less favorable circumstances or left unattended for longer time periods, lead to a severe consequence. If there is indeed potential for the observed anomaly failure mechanism to recur and lead to an accident (i.e. a situation that has more severe consequences), then the anomaly may be called an *accident precursor*.

> ***Anomaly Failure Mechanism (General Definition)***
> An anomaly failure mechanism is the principal underlying phenomena or behavior behind an anomaly, responsible for the observed off-nominal physical condition or operational behavior of the system.
>
> ***Anomaly***
> An anomaly is an off-nominal occurrence or condition (e.g. a deviation outside of certified or approved design or performance specifications).
>
> ***Accident Precursor***
> An accident precursor is an anomaly that signals the potential for more severe consequences that may occur in the future, due to causes that are discernible from its occurrence today. Such an event provides evidence that a failure mechanism is operative in the system and may pose a significant degree of risk, given the potential for it to recur with greater magnitude, or under less favorable conditions.

Based on the above definition, we may now recognize well-known examples of accident precursors:

- O-ring blow-by at Space Shuttle Solid Rocket Booster (SRB) joint locations, prior to the loss of the Challenger. On several occasions prior to the fatal Challenger accident, blow-by events were witnessed at SRB field joints. Based on available proceedings, there were discussions and meetings after the blow-by occurrences on the potential for greater consequences. The conclusions based on available knowledge were that there was no potential for significant consequences. This observation signified that a failure mechanism was operative in the system however the potential for severe consequences was misunderstood.

- Foam loss from the Space Shuttle ET and Space Shuttle TPS debris damage, prior to the loss of Columbia. On numerous flights prior to the Columbia accident, foam was observed shedding from the ET and impacting the Orbiter TPS. On several of these occasions the TPS tile was impacted and damaged, but never with catastrophic results. Most notably, STS-45 demonstrated that impacts were possible to the wing-leading edge Reinforced Carbon-Carbon (RCC) TPS panels, and that the RCC material could be damaged by the impacts, as shown in Figure 2-2. These events demonstrated that a failure mechanism leading to TPS damage was active on multiple flights, representing a potential for recurrence with a greater magnitude.

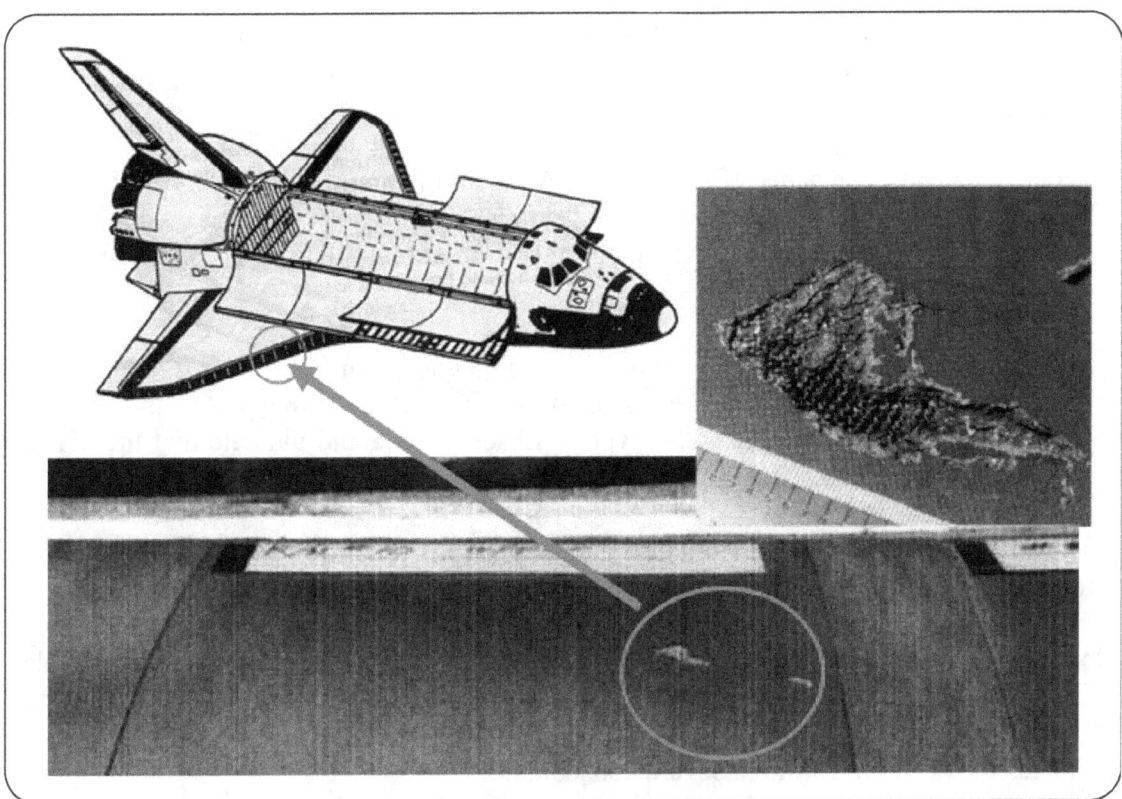

**Figure 2-2 Post-flight Inspection Photos of TPS Damage on STS-45**

A well-known example of a precursor in the nuclear industry was the increased rate of containment air filter clogging prior to the discovery of significant vessel head erosion at the Davis-Besse nuclear power plant [13]. Upon initial review it was believed that the cause of the air filter clogs was due to the filter itself and other anomalies not associated with the eroding vessel head. It was only after the discovery of vessel head erosion that plant personnel understood that the anomalous air filter performance was due to airborne material from the eroding vessel head.

Some examples of accident precursor types are:

- A near-miss because of chance or an opportune mitigation. An example of this type of precursor is the Shuttle TPS debris damage as observed on numerous flights prior to the loss of Columbia. On all previous flights critical TPS damage did not occur simply by chance that a debris impact of a great enough magnitude did not occur in a sensitive location.

- Faults that can become failure conditions without correction. An example of this might be a hairline crack in a fitting which is so small that it causes no leak or no loss of component function, but given time and use can grow to the point of leak or rupture.

- Unexpected operational behavior. For example, at times the operational environment of space can cause unintended effects to system operation. An example might be a lubricant which becomes more viscous than expected when operating in the temperature and pressure extremes of space and creates a threat to system function.

- Reduced maintenance effectiveness. An example might be a quality inspection of a system which over time becomes routine and possibly mundane such that gaps in the inspection develop which allow potentially harmful conditions to be accepted for flight.

- Unexpected effects from aging of equipment. An example of this type of precursor could be a coolant system where over time the pH of the chemical coolant drops as it ages. This type of observation could indicate that the pH will continue to drop and degrade cooling function.

The above examples of precursors and precursor types illustrates that there is no single template for describing an accident precursor. The connection between an anomaly and the potential for severe consequences can be relatively straightforward, as in the case of Columbia, or it can be indirect, as in the case of the Davis-Besse incident. It can relate solely to hardware behavior or it can involve human actions as well. The common element in all cases is an anomaly that is benign in its current instantiation, but which indicates the potential for more severe consequences.

## 2.2 Accident Precursor Analysis and its Role in System Safety

APA is the process by which an organization evaluates observed anomalies and determines if the mechanism at the origin of that anomaly could recur with more severe results. There is no one way to conduct APA, but all APA processes should evaluate operational experience to identify unrecognized accident potential or underappreciated vulnerabilities, so that something can be done about them *in a timely manner*[1]. APA is one analysis method that can be used in system safety to systematically incorporate operational data and experience in the overall safety analysis, allowing the prioritization of system and operational changes and improvements based on risk and potential risk reduction. Precursor analysis is important in that it 'teases' out information on 'emerging' safety issues using actual system operational experience to risk inform operational and system improvements.

The ability to assess the risk and safety implications of off-nominal system behavior is an essential part of an effective risk management and safety program, as recognized in NASA/SP-2010-580, NASA System Safety Handbook [6]. APA results are used to inform updates to the system's risk model; this feedback mechanism allows the real-world behavior of the system to be reflected back into the risk and safety analyses of the

---

[1] "In a timely manner" is a matter to be determined by the organization overseeing the system (as will be discussed in subsequent sections) but basically is defined by the end result – which is the avoidance of an accident due to a recurring failure mechanism.

system. In this way APA uses examples of off-nominal behavior in a *proactive* rather than a *reactive* fashion. The off nominal event is not simply resolved so that operation can continue; it is analyzed and used strategically by gleaning information from it to help understand and control risk for the future. As test and operational experience accumulates, the APA process helps to support a convergence between the assessed risk and its actual as-operated risk. In the absence of an APA process, convergence between a risk model and the occurring events and phenomena of the system modeled may occur in response to system failure, which for NASA systems is all too often catastrophic.

In order for NASA to conclude that a system is sufficiently safe, the information that demonstrates the system's ability to meet those levels of safety must be documented. This will consist of a consolidated set of technical and programmatic activities and standards that define and implement safety processes and requirements, and record operational performance, and system and operational changes. The culmination of that information is articulated in a case for system safety, or a *safety case*. An up-to date safety case will be expected to demonstrate that operational experience indicates that the system is as safe as desired and as claimed.

Figure 2-3 (adapted from [6]) illustrates the flow of information within a program in regards to real-world data and modeling efforts. On the left-hand side of the figure the real-world information is generated by the system and operational practices and is recorded in the form of off-nominal data reports (anomalies, non-conformances etc.). This data represents characteristics of the actual performance of the system. On the right-hand side safety and reliability analyses (Hazard Analysis, Probabilistic Risk Analysis [PRA] etc.) are created from performance modeling activities.

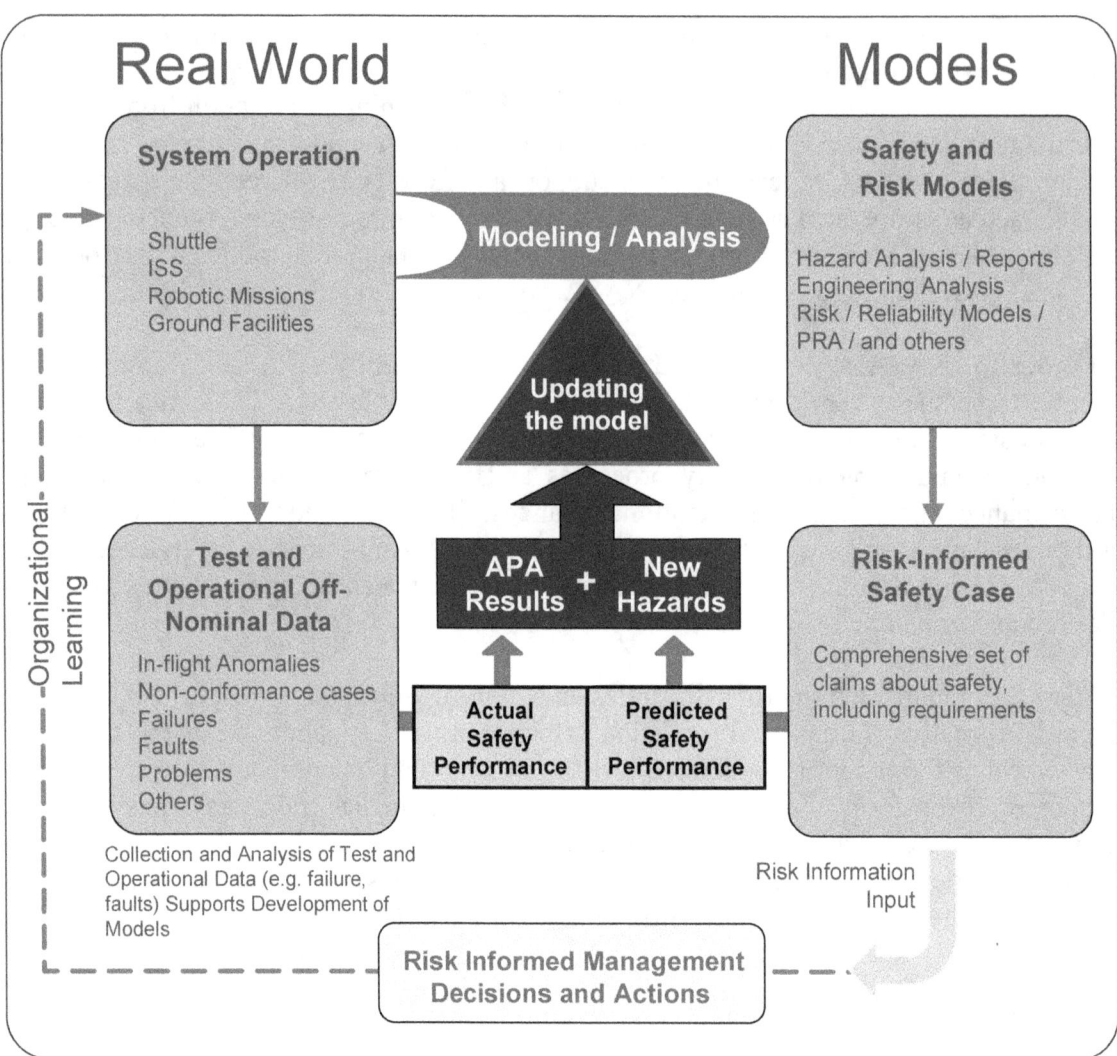

**Figure 2-3 Real World vs. Models Information Flow in the APA Context**

An understanding of system risk is developed by constructing the models illustrated in Figure 2-3. The structure of those models, as well as the data used to quantify them, initially reflects postulated system behavior informed by historical data, which does not necessarily correspond to actual system behavior. As the system is operated, APA uses operational off-nominal data to provide clues to its actual behavior (insofar as risk is concerned), which can be integrated into the model (potentially refining the structure as well as the quantification) in order to bring the predicted performance more in line with the actual performance.

Operation-informed refinement of the risk model to better represent reality is just one aspect of system safety. System safety as a whole depends on several elements including: measures taken to assure that systems are designed to effectively meet their intent, that the risks they present are identified and analyzed, measures taken to cost-effectively improve safety, and that operational procedures are in place to maintain the level of assessed safety. Taken together, these elements of system safety form the basis for the safety case, namely the information that NASA depends on for its conclusion that the system can be operated safely. Figure 2-4 (as provided in [6]) shows a notional representation of the safety claim model.

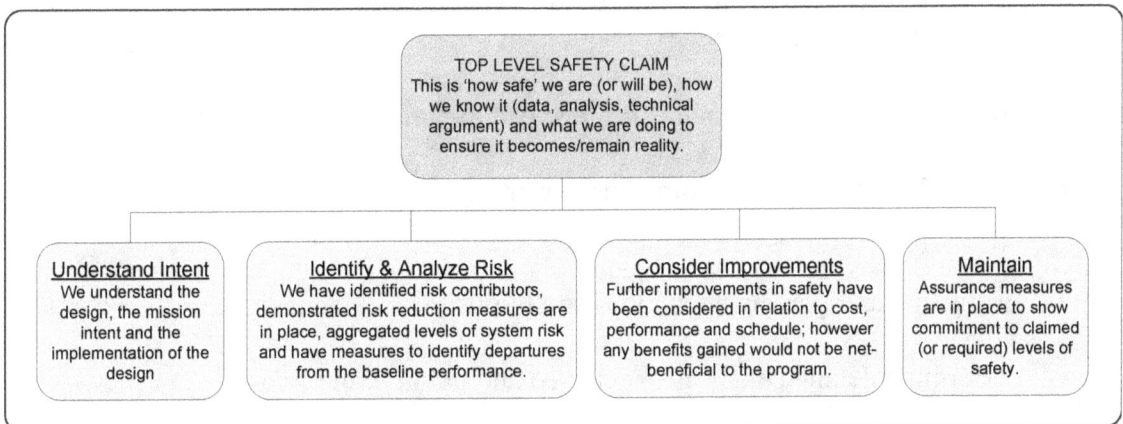

**Figure 2-4 System Safety Claim Model**

The underlying claims that ultimately provide confidence in meeting the requirements are a decomposition of the top-level safety claim. The first claim (*Understand Intent*) declares that the design, mission intent and implementation of the design (including procedures) are fully understood. APA includes a "Generalization" phase which, as will be discussed later, supports this claim by taking characteristics of known anomalies and applying them to other locations within the system and times within the mission; this highlights potential failure scenarios that may not have been considered in system design, and contributes to overall design understanding. The next claim (*Identify & Analyze Risk*) states that areas of risk have been identified, risk reduction measures have been implemented and that measures to identify departures from the baseline performance are in place. APA aids the identification and mitigation of risk by grading the anomalous conditions that represent the highest perceived risk potential and recommending them for risk modeling. If an under-appreciated risk is identified it drives programmatic actions to better understand the risk and implement risk reduction measures. The final claim (*Maintain*) states that the operator is committed to tracking safety performance and adhering to the levels of safety required. APA helps support the claim by identifying and prioritizing potential problems that might manifest in the future, based on what the system is demonstrating today. APA results are then used to update the system's risk model thus ensuring that the model is kept up-to date with actual operational occurrences of the system.

The safety case is maintained throughout the life of the system and matures throughout the lifetime of the program. APA is a continual process that analyzes the potential risk implications of anomalous events as they occur; a program of system safety that integrates APA demonstrates a commitment to safety assurance. To conclude that a system is acceptably safe it is fundamental that there be a process for determining the safety implications of off-nominal operational events or conditions, such as anomalies, and reflecting them back into the system's risk model.

## 2.3 NASA Accident Precursor Analysis Process Overview

APA establishes a systematic process for risk significance-based evaluation of operational and test anomalies by:

- Screening observed anomalies for the need to perform an evaluation

- Extrapolation of the anomalous event to other circumstances (anomalous conditions)

- Evaluating and grading anomalous conditions for further analysis

- Performing detailed analysis of selected anomalous conditions

Figure 2-5 presents a high-level view of the conceptual framework of NASA's APA approach, the process steps trace representative routes that an anomaly present in one of the "Anomaly Source Databases" (shown in the upper left-hand corner of the diagram) may follow as it makes its way through various screens, evaluation, grading, and analysis steps.

The process begins with a review of anomalous events as reported in existing databases (e.g., Problem Reporting and Corrective Action [PRACA] database), and a screening of those events that can be judged by either manual or automated inspection as having no practical relationship to any potentially risk-significant condition existing in the system of interest. This is necessary to focus the precursor analysis on events of most interest.

Events surviving this preliminary screen, meant to include all but the most clearly non-risk significant events (to minimize false negatives), are assessed for their causal failure mechanisms. These failure mechanisms are then generalized to different circumstances under which they might recur in the system, including the possibility of recurrence in different subsystems, at different times, or with different fault magnitudes. These generalized, postulated occurrences of the anomaly failure mechanism are referred to in APA as *anomalous conditions*, and characterize the potential for the failure mechanism that caused the observed anomaly to occur elsewhere in the system, with potentially more severe results. Note that Figure 2-5 depicts multiple outputs emanating from the "Generalization" step because a number of anomalous conditions may be postulated from a single anomaly as a result of the brainstorming activity that takes place during that exercise.

> **Anomalous Condition**
> In the context of APA an anomalous condition is defined as a *postulated* recurrence of an anomaly failure mechanism under circumstances that could result in more severe consequences than those that have been previously produced. Such circumstances include recurrence in different subsystems, at different times, or with different fault magnitudes.

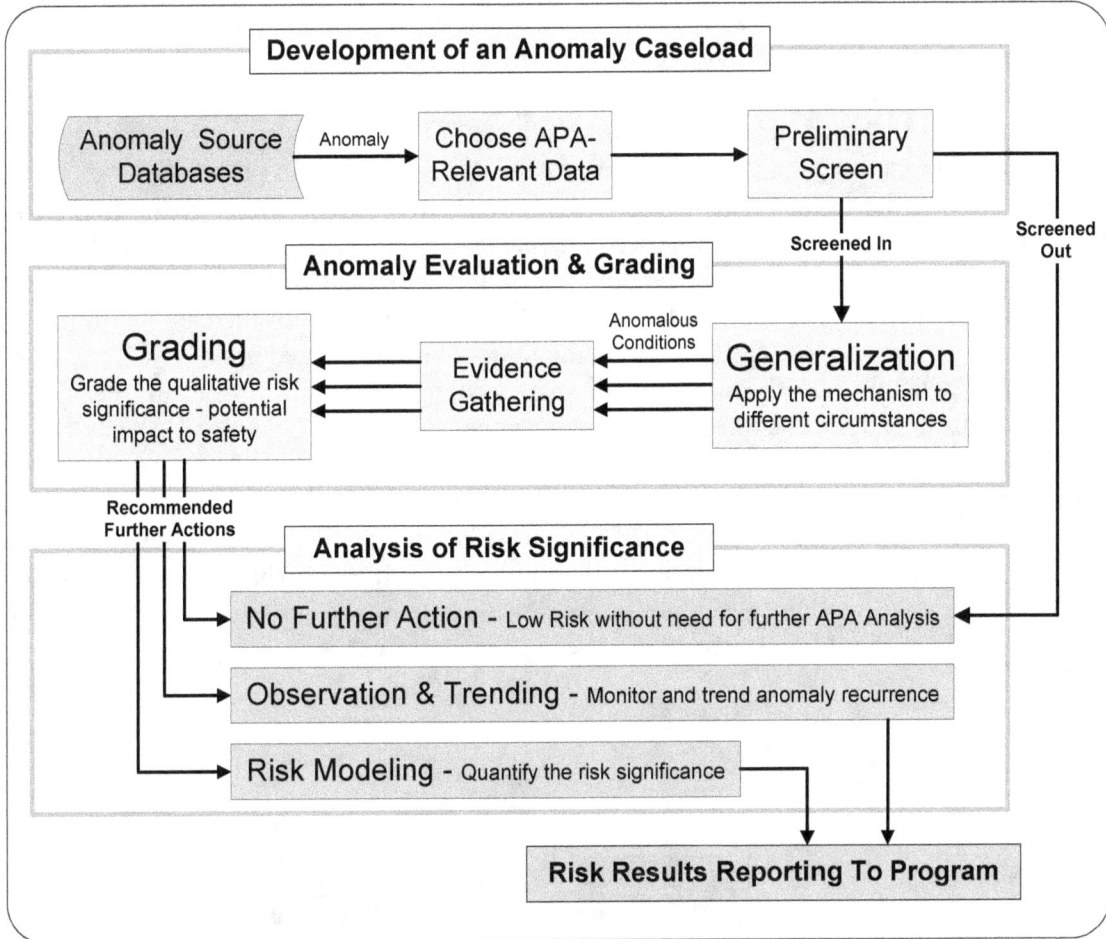

Figure 2-5 NASA Accident Precursor Analysis Process Overview Diagram

Once the Generalization process has been completed for a caseload of anomalies, information is gathered regarding the specifics of the anomalous conditions being pursued for further evaluation. This activity requires data-mining to investigate related engineering documentation (such as Failure Modes and Effects Analysis [FMEA], system schematics, physics models, test data, etc) and catalogue relevant data to be used as evidence in the following Grading exercise.

In order to prioritize anomalous conditions' in terms of their potential risk, a triage-like process called Grading is utilized. Grading is a judgment-based process applied to each anomalous condition to qualitatively assess their potential for producing severe consequences. For example, if an anomalous condition details an electrical short, Grading might estimate the potential for that short to result in a fire. This potential is characterized by a qualitative scoring of the potential risk called the *Potential Problem Index* (PPI). Anomalous conditions having a low assessed PPI are screened from further analysis. Anomalous conditions with an intermediate PPI are deemed to warrant continued observation and trending to assure that causally-related anomalies are identified, the underlying anomaly failure mechanism is characterized, and that effective controls are in place as needed to prevent the failure mechanism from propagating to severe consequences. Anomalous conditions with a high PPI are graded for future evaluation using scenario-based risk modeling in order to characterize the overall risk significance of the underlying failure mechanism.

Scenario-based risk modeling [6] may be conducted in the context of an existing system risk model, but can expand on an existing model using parametric probabilistic methods which associate the failure mechanism with the proper failure mode(s) in the model. In cases where the anomaly failure mechanism and the associated failure mode(s) are already adequately modeled, the anomaly provides no new risk information except perhaps as an incremental update to the statistical frequency of occurrence. Ultimately, when the APA process identifies a potentially under-appreciated risk it will catalyze action that will improve the understanding of system risk and modify the design or operation of the system as appropriate to address the issue.

The above APA process steps are implemented either by individuals or by group forums. Figure 2-6 shows a representation of the process steps and details whether they are offline tasks, or performed in group forums. Additionally, Table 2.1 shows these steps and identifies the skill set that the individual performing them should have or that the team performing them should collectively possess.

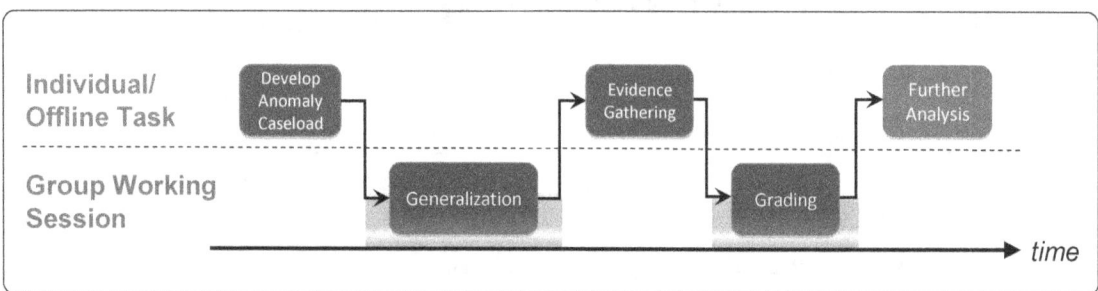

**Figure 2-6 Timeline of APA Process Steps**

The NASA APA process may be applied as soon as there are anomalous events that will impinge upon the current or impending operation of a particular system. In some cases this may be as early as the Preliminary Design Review (PDR) in the development portion of the system life cycle. For instance, if the design solution at PDR includes heritage equipment from previous NASA programs/projects or even external applications of that equipment, the anomalies that have historically occurred on that equipment can be evaluated to determine what, if any, risks they portend for the system being designed. This is already done to an extent by incorporating lessons learned from historical systems; however APA provides the added benefit of generalization which helps to extrapolate those lessons learned onto other systems or other scenarios. Once prototype subsystems begin testing, usually between PDR and the Critical Design Review (CDR), test anomalies can be evaluated to both inform the FMEA as well as integrating the results of the APA into system risk models. The combined findings from APA and FMEA can, of course, be utilized to effect design changes that may eliminate or mitigate unacceptable risks early in the design process when the cost of making those changes is relatively low. After the system begins operation, it is prudent to continue applying the APA process to root out under-appreciated risks and to identify new risks that crop up as the system's processes, missions, and operating environments change, or in the case of reusable systems, as they simply begin to show signs of degradation due to aging.

The APA process utilizes a graded approach, i.e., the level of effort and detail of the analysis is commensurate with the current state of knowledge regarding the risk significance of the failure mechanism being analyzed. It leverages existing risk analyses, using elements as appropriate and filling in (or, at a minimum, identifying) risk-significant information gaps. This is accomplished by applying a standard of evidence to both qualitative and quantitative assertions relating to the potential of the failure mechanism to lead to a severe system consequence, identifying areas of uncertainty

Table 2.1 NASA APA Process Steps and Required Skill Sets

| APA Process Step | Individual or Team Activity | Skills or Expertise Required |
|---|---|---|
| Choose APA Relevant Data | Individual | • Familiarity with anomaly reporting system<br>• Competency in data mining and data interpretation<br>• Moderate engineering knowledge of the respective program |
| Preliminary Screen for Non-significant Events | | This is a process step that can be implemented once the subsequent teams define the screening criteria |
| Generalization | Team | • Familiarity with anomaly reporting system<br>• Accident Precursor Analysis expertise<br>• Coordination of technical discussion group skills and leadership expertise<br>• Ability to concisely transcribe the decision and rationale of the group discussion<br>• Systems engineering experience; comprehensive knowledge of system function, operation and system integration |
| Evidence Gathering | Individual | • Competency in data mining and data interpretation<br>• Moderate technical knowledge in the relevant engineering field<br>• Support provided by subsystem experts as required. |
| Grading | Team | • Accident Precursor Analysis expertise<br>• Coordination of technical discussion group skills and leadership expertise<br>• Ability to transcribe the decision and rationale of the discussion<br>• Experienced systems engineering skills - comprehensive knowledge of system function, operation and system integration<br>• Detailed subsystem expertise; function, operation, failure modes & effects, and integration with related subsystems |
| Risk Modeling | Individual | • Systems engineering and probabilistic risk modeling expertise<br>• Support from subsystem experts as required. |
| Observation & Trending | Individual | • Data interpretation skills<br>• Moderate engineering experience in the respective program<br>• Support from Systems Engineers. |
| Risk Level Reporting | Individual | •Expertise in the program's 'Safety, Reliability and Quality Assurance' processes<br>•Knowledge of the program's decision making protocols and reporting requirements" |

where additional analysis, inspection and/or testing may be warranted in order to confidently understand the failure mechanism and the associated response of the system. This evidentiary standard also applies to existing analyses, to ensure that risk results are empirically supported, either by direct statistically significant testing or by the use of appropriately baselined physics-based analysis.

For those anomalous conditions that are graded for risk modeling, the main intent is to estimate the risk of severe consequences, given an occurrence of the anomalous condition, and to identify the drivers of that risk in terms of the physical conditions that produce it. This is done by conservatively characterizing the uncertainty in the parameters of the system risk model, and determining the parameter values that produce the highest risk. This enables decision makers to focus resources to most effectively reduce the potential for the system to operate with these values, or to assure that operation under such conditions does not present undue risk.

In summary, the basic principles of the NASA APA process are:

- **Anomaly data may point to anomaly failure mechanisms that, under different circumstances, could result in severe consequences.** The prudent response to such anomalies is to assess this potential to assure that no unknown or underappreciated risks exist in the system. The APA process generalizes from the particulars of the observed anomaly to the spectrum of potential failures that share the same causal mechanism but which may occur in more severe form, at different times, or in different locations.

- **Use of a common risk measure.** APA provides a broad review of all anomalies and utilizes a common risk measure over the full range of anomalies. This assures consistency in the management and disposition of risk issues arising from the APA process.

- **An efficient APA process depends on a graded approach.** Anomalies are assessed by screening, qualitative assessment and quantitative modeling. Quantitative modeling is used only as needed to understand the system risk attributable to the failure mechanism.

- **The process is based in evidentiary support.** Anomaly failure mechanism generalization involves the postulation of fault magnitudes, propagation pathways and system stresses that may not have occurred within the operational experience base of the system. Thus, uncertainties are potentially large and care must be taken to avoid mischaracterization. The APA process explicitly weighs the evidence that is used to support the characterization of risk, implying where additional testing or analysis may be beneficial.

- **The focus is on identification of underappreciated risks.** Anomalies are significant to the extent that they either imply performance degradation of well-understood elements or reveal risks that were underappreciated or misunderstood prior to the anomaly occurrence. APA makes use of existing risk models to ascertain the risk significance of the failure mechanism, and entails additional risk modeling as needed to understand the physical drivers of that risk.

- **The APA process is designed to produce actionable findings.** It is expected that the findings will be available to the risk management system to support the development of risk-informed recommendations for system operation and testing, as well as assessing the adequacy of any corrective actions that may already have been taken in response to the original anomaly. Therefore, top-level metrics that measure the risk importance of the underlying anomaly failure mechanism, as well as the associated physical parameters that may contribute to a catastrophic exacerbation of that mechanism, are fundamental to the process.

# 3 Accident Precursor Analysis Process Steps

## 3.1 Building a Caseload

The APA process begins with the compilation of an anomaly caseload, collected from reports of anomalous events which occurred within the system under evaluation. It is these reports which are used as the initial indicators by which precursor analysis is conducted and underappreciated risks may be identified. At a minimum, the caseload should be a set of actual events that occurred during system operation that reveal an anomaly in the physical characteristics or behavior of the system. Within large systems such as the Space Shuttle or ISS, a correspondingly large number of anomalous event reports or event reporting systems may exist. In this case, it is the job of the precursor analysis team to assemble all of the reported data, filter out that which is not of value to the analysis process, and leave a refined, manageable set of significant and valid data to evaluate.

### 3.1.1 Data Sources of Interest

Typically, anomalous events are reported through a problem reporting process where a technician or mission operator will observe an anomaly and provide a description of the event. Information regarding the anomalous event is then entered into a database to be used at a future date. Event reports for use with APA should at a minimum contain a short title and a description of the anomaly as it was observed, time and date information of the anomaly occurrence, as well as some descriptive element which sheds light onto the causal failure mechanism of the anomaly. Typically much more information will also be available, some of which may help to adequately characterize and classify the event that was observed. Some common types of information which may be recorded in an event report are the system and subsystem in which the anomaly was observed, engineering contacts responsible for the system, remaining controls against potential failures and the potential consequences of such failures (by referencing a FMEA for instance), and links to outside documentation on the component(s) that experienced the anomaly. For the purposes of APA, it is important to choose anomaly data sources that contain a maximum of information relevant to the identification of anomaly failure mechanisms, controls preventing system failure, and potential consequences of failure.

### 3.1.2 Choosing Accident Precursor Analysis Relevant Data

Since the APA process is concerned with finding underappreciated risks which could result in severe consequences, it is important for the anomaly data source to minimize the inclusion of reports documenting conditions of negligible risk. For example, anomalies discovered during acceptance-testing prior to system operation document non-conformances within operationally-certified systems, but represent the testing system at work, preventing the non-conformances from being accepted for nominal use where they may cause failure. Conversely, records which document events reported during system operation demonstrate failure mechanisms which passed through the rigors of acceptance testing, and thus demonstrate discernable potential threats to safe system operation during flight or system operation. Additionally, events reported during testing of a design that is approaching its final design configuration can demonstrate the ability for failure mechanisms to manifest themselves in flight-like hardware, and should be viewed as potential threats to safe operation. For this reason, it is recommended that the anomaly data pulled for APA be limited to anomalies that occur on systems approved for operation or operation-like testing. If time and resources allow, further examination of anomalies cataloged outside of system operation (i.e. acceptance testing, fleet leader failures, etc.) could be considered as well.

> *Space Shuttle Example*
>
> In the Space Shuttle Program, anomalies are recorded in different repositories or with different designations and resolved in accordance with different processes, as appropriate to the specific anomaly, such as in the Space Shuttle PRACA system. PRACA-reportable anomalies occurring on flight hardware or software during a mission phase of the system lifecycle that meet certain criteria for immediate attention are designated as in-flight anomalies (IFAs) and their disposition is reviewed by the Program Requirements Control Board (PRCB) prior to the next flight. This is in contrast to PRACA-reportable anomalies as a whole, which tend to be of a more test-oriented and maintenance-oriented nature and do not necessarily warrant an in-depth assessment prior to the next flight. Generally, the information recorded in the PRACA system allows areas in need of correction or improvement to be remanded to engineering for development of a corrective action, if deemed necessary.

If multiple data sources are available that contain records of unique anomalies, then a combined data set may be created and used as a dedicated APA database. Equivalent or very similar fields from the original databases can be grouped together for the APA database, but unique fields from each database should be preserved in the combined set, ensuring that the combined database contains all information from the original sources. The diagram in the following Space Shuttle example shows this concept in practice, when fields from two databases were combined to make a comprehensive database. The end objective of this step is to establish a single list of anomalies for the deliberative process

> *Space Shuttle Example*
>
> In the space shuttle program, two valuable databases exist which exemplify the need to combine databases in some cases. One is the aforementioned PRACA IFA database. Another example of such a source are the In-Flight Anomalies (IFAs) recorded by the Mission Evaluation Room (MER). The MER is a safety and mission assurance board which provides real-time support for missions. If any anomalous condition or non-conformance is reported on the space shuttle, an operator on one of the MER consoles records the issue, generating a report for immediate investigation by the board regarding its potential safety and risk implications to the specific mission taking place.
>
>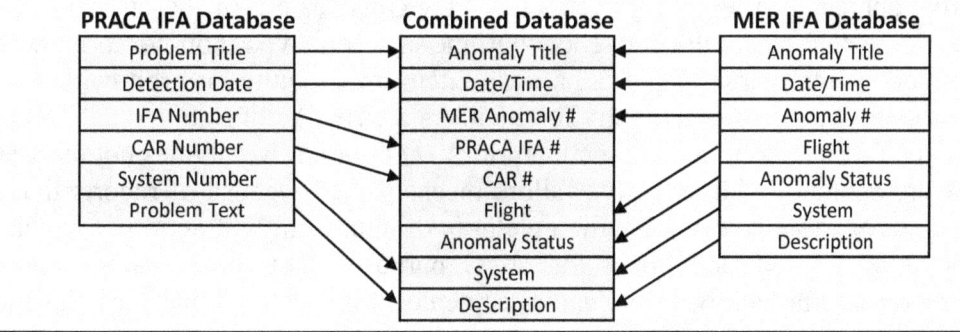

### 3.1.3 Timetable for Evaluation of Anomaly Data

Often when anomaly reports are initiated only some core information is recorded and the rest is filled in as information is gained about the condition (e.g. from a root-cause analysis) or after a corrective action is put into place. Accordingly, consideration should be given to the time at which the anomaly source data will provide the most value to the APA process. For example, given the process's emphasis on generalizing from the causal mechanism, it is valuable to wait for a causal analysis to be performed on the anomalous event if one is indeed planned. It will not always be possible to wait for a causal analysis to be completed on all anomalies in a data set; however, it is suggested that any steps that improve the understanding of the underlying failure mechanism be accomplished before the anomaly is evaluated as part of the APA process. It should also be understood that there is a risk associated with any delay in applying APA, since it might extend the window of opportunity for the failure mechanism to reassert itself and result in severe consequences. In actual applications for both the Space Shuttle and ISS programs, the process worked well when performed approximately six months to one year after the most recent anomaly in the data set. Given that amount of separation time, few records lacked the investigative background information to proceed with the process, or the causal information to identify the anomaly failure mechanism. Ultimately, however, the decision of when to apply the process must be tailored to the specific program conditions, in order to balance the need for a more complete understanding of the risk-significance of the anomaly against the risk of the underlying failure mechanism recurring with more severe consequences.

## 3.1.4 Anomaly Screening Methods

The objective of an anomaly screening step is to filter out anomalies that obviously have little or no potential for safety impact. As discussed previously, anomaly reporting at NASA is carried out to serve diverse purposes, and as a result, a number of anomaly reports have little or no safety nexus. In addition, reporting is often implemented conservatively; in order to assure that no situation is disregarded that could potentially fall under the definition of an anomaly or nonconformance in accordance with NASA best practices [14], mission evaluation and problem reporting personnel typically take a conservative approach to reporting events that they consider out of the ordinary. Often, these are not actual anomalous events, but merely perceived non-conformances or misunderstood events. For example, a hardware item may require on-orbit calibration to function properly in a zero gravity environment. In a case like this an anomaly report is opened to document the erroneous function prior to calibration, yet it does not represent a true hardware anomaly with an active failure mechanism. Because of reports like this, problem reporting systems may contain a plethora of items, only a fraction of which will warrant a more detailed assessment for safety purposes. This conservative nature of anomaly reporting may allow for an initial screening for incidents that are relevant for APA. The screening process can be implemented by filtering the collection of anomalies based on a predefined set of screening rules. The screen should be quick and relatively effective at trimming off events whose causal mechanisms have no possible impact on safety or system reliability. It should not, however, inadvertently remove a potential precursor event from further consideration. Therefore, false positives (over-conservative assessments) are to be tolerated and expected in this preliminary screening of reported anomalies, while false negatives (non-conservative assessments) are to be minimized.

The rules by which one screens anomalies are by no means standardized or are expected to be the same across different programs. Ideally, the application of the APA process to a new program will start without a screen, and gradually develop the rules for one as early applications of the process are conducted. The rules for screening are very dependant on context, and some programs may have no screening procedures at all, while others may have many.

> *Space Shuttle Example*
>
> In the Space Shuttle MER IFA database, the anomaly numbering scheme has one prefix for some anomalies, and a different prefix for others. Investigation into this phenomenon with MER personnel revealed that all records, when they are recorded, are given one prefix. Later, when they are further investigated and deemed to be an actual anomalous event, the prefix is changed to a different one. Therefore, a screen that filtered for the prefix which denotes an anomaly was established thereby "screening-in" only true non-conformance events.

Additionally, in some cases systems or types of hardware as a whole may not be capable of either creating a severe consequence or possessing an anomaly failure mechanism which could manifest itself in other systems. For example, soft goods or crew personal effects may be phenomenologically benign or functionally not necessary for a successful flight. This situation, if it occurs, will likely apply to either very unique or very isolated systems, for which postulating failures from the same mechanism, but outside of system borders, is not reasonable. If such cases exist, and the system as a whole can be deemed as non-critical to reliability or safety, then they may be screened out.

When creating an anomaly screen, one should be careful to make a distinction between "non-anomalies," and "non-critical anomalies." Some event reports may document an anomaly; however, they may note that the anomaly was not critical to system reliability or safety. APA is focused upon identifying the causal failure mechanisms which initiate anomalies, and postulates other locations or times in which the same mechanism could have more severe results. As such, an anomaly report which identifies any failure mechanism, no matter how benign in the observed anomaly, is of potential value to the process as an indicator of an underlying failure mechanism that could have more severe consequences at other locations in the system or at other times in the mission. Any screen therefore should not "screen-out" anomalies simply based on the recorded criticality for that specific anomaly.

## 3.2 Anomaly Failure Mechanism Identification & Generalization

In order to assess the risk significance of the anomalous events in the anomaly caseload, it is first necessary to differentiate between that part of each anomaly that is characteristic of the system design and/or operation (and thus potentially recurring), versus that part that is circumstantial to the specific anomaly instance that actually occurred. This enables the APA process to evaluate the ways in which the characteristic aspect of the anomaly might recur, under less favorable circumstances that could lead to adverse consequences. The process step for doing this is called *generalization*[2].

Generalization instills within the APA process the ability to go beyond the circumstantial aspects of the anomaly *as it occurred*, to the spectrum of possible instantiations of the causal failure mechanism, i.e., *what might occur*. The scope of generalization goes beyond the immediate issue of assessing and mitigating the anomaly itself. Instead, it is focused on the anomaly's causal failure mechanism as it might arise in other circumstances, in order to identify design or operational vulnerabilities to the mechanism in a broader sense. Generalization produces a set of anomalous conditions which characterize the potential for the underlying mechanism to occur in the system as a whole. Anomaly generalization is illustrated in Figure 3-1, with the expanding cone representing the increasing number of postulated anomalous conditions originating from the initial anomaly review.

---

[2] The label "extrapolation" has also been suggested to emphasize the point that during this step observed phenomena are considered in different contexts.

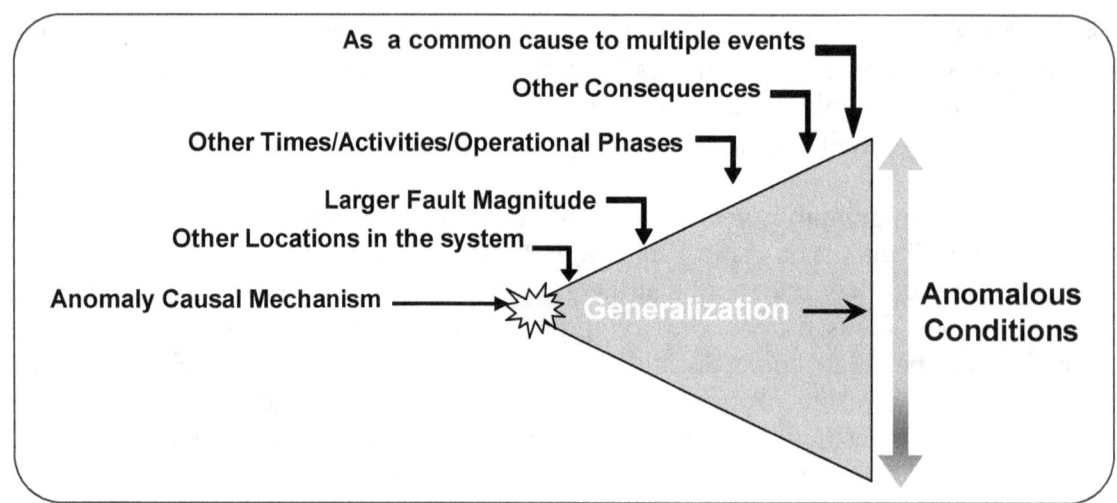

**Figure 3-1 Anomaly Generalization Illustration**

### 3.2.1 Identifying the Anomaly Failure Mechanism

Ideally, the timing of the APA process relative to other anomaly investigation activities will be such that a causal analysis has been performed and is available as the basis of generalization. When this is the case, the causal analysis is reviewed to determine the appropriate point in the causal chain at which to define the failure mechanism. This point is not necessarily at the level of the "root cause," defined by NASA as an organizational factor that contributes to or creates a subsequent undesired outcome [15]. Generalizing from organizational or root causes (e.g., failure to understand the importance of training) would tend to result in an unmanageable number of anomalous conditions, most of which would bear little, if any, phenomenological resemblance to the observed anomaly. Conversely, proximate causes (e.g., pipe leak) tend to be the result of prior physical conditions (e.g. improper installation, fitting wear-out, or contamination of fitting). Generalizing from proximate causes therefore tends not to inform the development of implementable recommendations to reduce system vulnerability. Instead, as a rule of thumb, the causal failure mechanism used for generalization should be the first physical contributor to the off-nominal condition of the system. In the above example of a pipe leak as a proximate cause, the intermediate causes of improper installation, fitting wear-out, or contamination of fitting would each be a reasonable anomaly failure mechanism from which to generalize, since they represent the initial (or at least early) *physical* consequences of anomalous events (e.g. other examples include installation error, debris impact, electrical short and inadequate maintenance). In this way the identification of the anomaly failure mechanism is somewhat subjective. For any given anomaly there may be several ways to define the anomaly failure mechanism; this is acceptable within the APA process. Also note that this example shows that both phenomenological (wear-out, contamination) and procedural[3] (improper installation) failure mechanisms are

---

[3] Procedural errors can include those made during system operation as well as processing errors made during hardware assembly prior to operation

appropriate for use with the APA process. The process itself focuses on risks to safe operation during flight, but the origins of those risks can occur prior to flight (assembly, ground processing, etc.).

> **Failure Mechanism (Accident Precursor Analysis Operational Definition)**
> For the purpose of APA, the anomaly failure mechanism is the initial off-nominal physical condition in the system and the jumping-off point for anomaly generalization. Examples of possible anomaly failure mechanisms include improper installation, fitting wear-out, and contamination.

In the case where a causal analysis has not been conducted, a decision must be made whether there is sufficient information, such as from the analysis of similar anomalies, to reasonably determine the anomaly's causal mechanism. If the possible causes of an anomaly can be reasonably narrowed down to a candidate failure mechanism this can be used as the basis for generalization, despite the possibility of investigating a mechanism that was not actually operative for the anomaly. Otherwise, if no reasonable failure mechanism can be established, this fact is noted and generalization of the anomaly is postponed pending the identification of a causal mechanism.

## 3.2.2 The Scope of Generalization

The intent behind Generalization is to identify design and operational vulnerabilities due to the failure mechanism. To limit the process of Generalization, it is restricted as a rule to the type of component(s) that are at risk to this mechanism. Generalizing the failure mechanism to other locations within the system is constrained to a *generic component type* that experienced the observed anomaly. Thus, for example, if the anomaly was a slow opening regulator valve, the generic component type might be regulator valves; the scope of generalization would then be limited to other regulator valves in the system, as opposed to being generalized to all valves. If the anomaly involves a leak resulting from a particular failure mechanism, then the population of similar components that are susceptible to that mechanism will be examined. If the anomaly involves cracks, then the susceptible population is the similar components that are made of similar material and subjected to similar stressors. The intent is to systematically expand the evaluation of the original anomaly to consider components of the same design or function that may also be susceptible to the same failure mechanism indicated by the anomaly event.

In some cases the anomaly failure mechanism or other characteristics of the anomaly may make the scope of generalization very large or very small. For example, consider an anomaly which was caused by crew procedural error. Procedures may be written correctly, and the crew may be adequately trained for the procedure but the human-element always allows for human error. In this case, generalizing to other areas in which human error could create a problem would be seemingly endless, and thus the scope of generalization can be deemed too broad. On the contrary to this, some anomalies may be the result of a failure mechanism so specific that it cannot be generalized to other components. An example of this may be a software error that is attributed to one unique string of code, such that the error couldn't possibly occur in other pieces of software. In both of these cases generalization is not practical, and the generalization of the anomaly can stop.

### 3.2.3 The Generalization Group

During the APA generalization activity the team works through a series of anomaly reports (the anomaly caseload) one-by one, creating suitable anomalous conditions. In practice the team often encounters an anomaly that is actually a duplicate or recurrence of an anomaly previously generalized. To streamline the process, the duplicate or recurrent anomaly is assigned to the same *generalization group* as the original anomaly. The APA tool, DAnGERS, has been designed to assist in this step by searching for and identifying items which may belong to common generalization groups.

Generalization groups are a useful way of grouping together similar anomalies (and consequently, also the resulting anomalous conditions). A generalization group gathers together observed anomalies and the postulated anomalous conditions composing of the same failure mechanism acting upon the same component type, i.e. everything within the same *scope of generalization*.

Organizing anomalies into generalization groups provides two advantages;

- It prevents the generalization team from generalizing the same type of anomaly twice. NASA anomaly reporting databases typically contain anomalous events of similar characteristics; a number of anomalies in an APA caseload often consist of the same failure mechanism acting upon the same generic component type. Even with two different anomalies one can create a matching group of anomalous conditions, if they consist of the same failure mechanisms acting upon the same generic component type. There is a many-to-many relationship between anomalies and anomalous conditions within a generalization group (see Figure 3-2), as opposed to the one-to-many relationship that would be expected from an isolated anomaly. Ideally, the first anomaly of a generalization group will be used, via generalization, to scope out the set of anomalous conditions that span the generalization group. Thus, when subsequent anomalies in the generalization group occur, they are not expected to produce additional anomalous conditions unless some were unidentified previously. This practice of organizing anomalies into generalization groups saves time and effort during generalization sessions since it prevents the team from repeating deliberations and creating duplicate copies of anomalous conditions previously generalized.

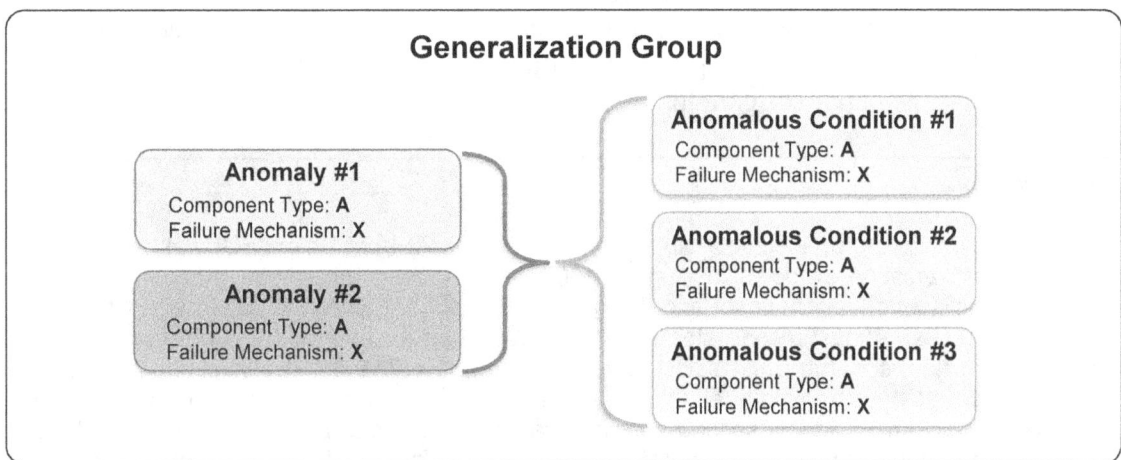

Figure 3-2 Many-to-many Relationship between Anomalies and Anomalous Conditions

- It provides a platform for grouping together related anomalies, for the purposes of future trending analysis. Within the context of APA, recurring anomalies consist of the same failure mechanism acting upon the same component type; therefore all the observed anomalies within a generalization group can be considered to be a series of recurring anomalies. This concept is presented in Figure 3-3; where a succession of anomalies occur along a timeline, each one consisting of a certain failure mechanism (denoted A,B,C) and a generic component type (X,Y,Z). The anomalies are organized into generalization groups according to their failure mechanism and generic component type; those composed of matching component type and failure mechanisms are classified as recurring anomalies and segregated into their respective groups.

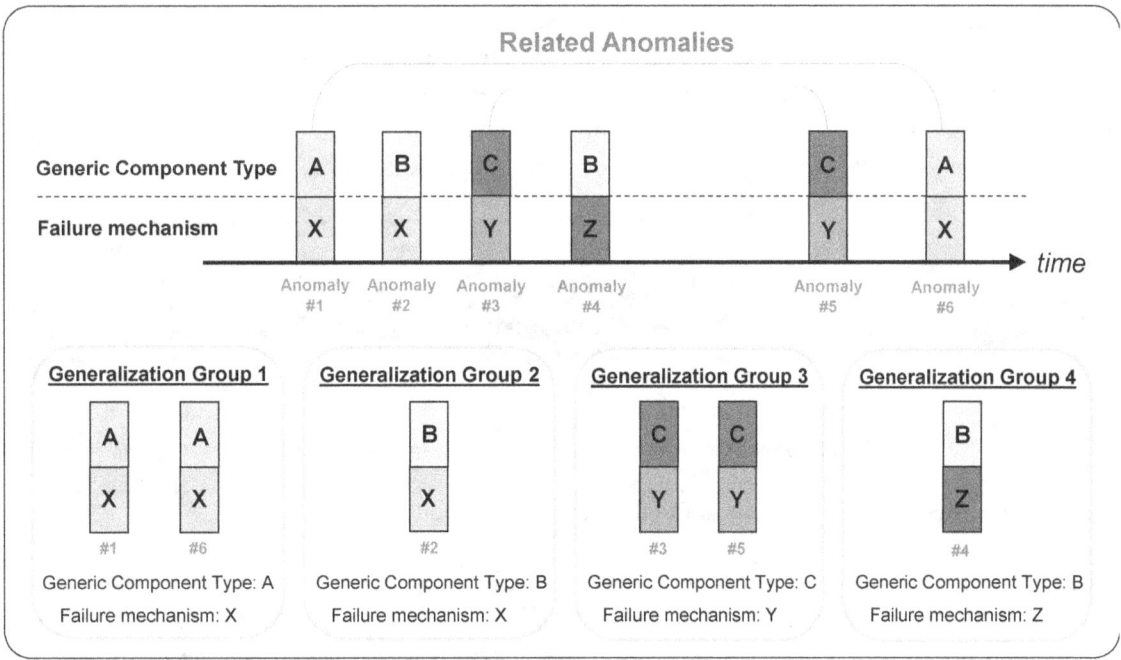

Figure 3-3 Recurring Anomalies in Generalization Groups

Parsing anomalies into generalization groups enables the APA process to actively group together related anomalies during generalization, thus creating suitable populations for the purposes of potential Observation & Trending (See Section 3.5 for further details).

> *ISS Example*
>
> An anomaly involving an electrical fault within a Sequential Shunt Unit (SSU) was caused by high-energy radiation particles in August 2009. In this case the failure mechanism was classified as a *single-event upset* and the generic component type was *External Local Data Interfaces (LDIs)*. Almost a year later the same subsystem experienced an anomaly within a Battery Charge/Discharge Unit (BCDU) involving an electrical fault. After investigation it was discovered that the anomalous event was caused by high-energy radiation particles. The failure mechanism within this second anomaly was identified as a *single-event upset* and the generic component type was *External Local Data Interfaces (LDIs)*. Although these two separate anomalies occurred within different locations and at different times, they are intrinsically related and tell us something valuable about the occurrence of high-energy radiation in external LDI components. If a trending activity were to be recommended for either of these anomalies, both events should be included as data points within the same trend.

See Appendix C for DAnGERS illustrations on creating and managing generalization groups.

### 3.2.4 Anatomy of the Anomalous Condition Accident Sequence

As noted, at the onset of APA is the identification of a failure mechanism from which a set of generalized Anomalous Conditions are formulated. Each of these anomalous conditions is used to define an accident sequence scenario by which the failure mechanism could ultimately progress to a severe consequence. The structure of the accident sequence is defined by three points as shown in Figure 3-4 and will become the backbone for the structured APA review of anomalous conditions.

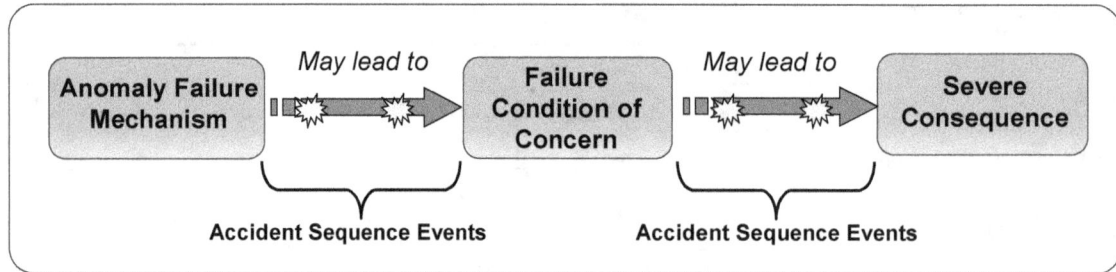

**Figure 3-4 Structure of an Anomalous Condition Accident Sequence**

The accident sequence for each anomalous condition is defined as the Anomaly Failure Mechanism progressing to an intermediate failure state called the Failure Condition of Concern, and then that failure state ultimately progressing to the severe consequence. The accident's progression between each of these states may involve many different events along the accident sequence, all of which will help to characterize the specific accident sequence under consideration. Each of these terms is defined further in Section 3.2.5.

## 3.2.5 Generalizing the Anomaly Failure Mechanism

Once the anomaly failure mechanism is identified, it is generalized throughout the system. The deliberative process considers the potential recurrence of the anomaly causal mechanism with varying characteristics in the following fields:

- **Fault Magnitude** – Anomalous events typically manifest in a range of fault magnitudes, with the majority of events lying at the smaller end of the magnitude spectrum (many do not even result in a fault). Generalization also considers the potential for the failure mechanism to produce a higher fault magnitude than was previously experienced by the anomaly. The presence of a larger fault magnitude in the system can present a higher potential for severe consequences; e.g. a crack in a fluid line may result in a release of fluid at some leak rate. A high leak rate caused by a large crack may result in a sufficient volume of fluid that presents a toxic/fire hazard.

- **Susceptible Locations** – Each component of the identified type is evaluated to identify locations that are susceptible to the failure mechanism, which could potentially lead to failure, and that may have the potential to lead to severe consequences. If a component is on a system whose failure clearly has no safety implications, and which is sufficiently isolated from other systems that failures are locally contained, then the component would not be considered susceptible, since there is no potential for severe consequences. This is the case even if the component is physically susceptible to the failure mechanism. Only when a component is physically susceptible to the failure mechanism *and* there may be a potential for severe consequences is the location considered susceptible from the perspective of APA.

  The goal at this point is to determine the scope of susceptible locations that could be forwarded to the grading exercise, making efforts to identify all locations where the anomaly failure mechanism has the potential to lead to severe consequences, while also minimizing the number of ultimately insignificant anomalous conditions.

- **Exposure Time/Activity** – NASA systems are typically dynamic, changing through numerous flight segments, environments, configurations, and operational states or phases. Failure mechanisms that occur benignly at one point in time might be catastrophic in another, even for the same location (i.e. an external ammonia fluid line leak during normal operations might not result in any severe consequences; however if the same leak were to occur during Extravehicular Activity [EVA] maintenance, ammonia fluid could collect on an Extravehicular Mobility Unit [EMU] resulting in potential atmospheric poisoning upon EVA ingress). Therefore, to provide an operational context for subsequent assessment, each anomalous condition is qualitatively evaluated to identify the time period or activity during which the failure mechanism could have critical results. This entails a degree of system-level evaluation of the possible effects of the failure condition of concern, but the intent is not to conduct a detailed investigation into its possible propagation pathways. If qualitative evaluation suggests that the system may be vulnerable to the failure condition of concern during multiple periods (e.g., during distinct mission activities), then each should be separately captured.

- **Failure Condition of Concern** - For components that are susceptible to the failure mechanism, the Generalization Team (see section 3.2.6 for details) identifies the *failure condition of concern*, i.e. the failure state that could be caused by the failure mechanism, and which has the potential to propagate to severe consequences. The failure condition of concern will typically correspond to a genuine failure of a particular item or function, as opposed to a merely out of spec condition. It may vary from location to location, even for the same failure mechanism, due to the different functions and system interactions associated with different components. For phenomenological failures, the failure condition of concern will typically represent a hazardous environment of a sufficient magnitude to represent a threat to the neighboring equipment. Thus, for example, if the failure of the susceptible component is a hydrazine leak, then the failure condition of concern would indicate, at least qualitatively, the size of leak necessary for producing a local hazard.

  The failure condition of concern is a key element of the accident precursor analysis process. It structures the assessment of susceptibility by providing an anchor point for consideration of propagation pathways from failure mechanism to severe consequences. Given a failure condition of concern, the assessment can be partitioned into a portion from failure mechanism to failure condition of concern, and a portion from failure condition of concern to severe consequences (see Figure 3-4). Moreover, the failure condition of concern, and the partition it represents between the failure mechanism and the severe consequences, is fundamental to subsequent APA process steps. Identifying the failure condition of concern at this point in the process facilitates the data collection activity (discussed in the next section) that supports these subsequent steps. The failure condition of concern should be described to a level of detail that will allow a data miner to find all pieces of information that could be of potential benefit for assessing the actual potential for the failure mechanism to lead to the failure condition of concern, as well as gathering information related to the potential for that failure condition of concern to propagate to severe consequences.

- **Severe Consequence** – The 'brainstorming' activity of generalization occurs within the context of the severe consequence(s) defined by the program. A severe consequence serves as the terminal event in the accident sequence characterized by the anomalous condition (see Figure 3.4). Severe Consequence(s) are typically defined by the program's risk management practices e.g. Loss of Science, Loss of Mission, Loss of Crew, the terminal events of the generalized anomalous conditions within APA must correlate with these. Once the location, exposure time/activity, and failure condition of concern have been determined, the severe consequence that could potentially be realized is identified. For programs with a single severe consequence of interest there is no decision to be made and only those accident sequences that can credibly produce the single severe consequence are considered. For those programs which have multiple severe consequences defined, the scope of generalization is broadened to include anomalous conditions that may propagate to a variety of terminal events. In some cases, an anomalous condition may exhibit the potential to result in more than one severe consequence under different circumstances (example: an electrical short on a circuit board within a science rack may lead to loss of science; however the same electrical short on a circuit board whilst in the presence of a hydrazine leak may result in fire ignition and loss of crew). The APA process allows anomalous conditions to be generated for multiple severe consequences.

Engineering judgment, conservatively applied, is sufficient to identify a set of anomalous conditions. It is not meant to include every possible instantiation of the failure mechanism within the susceptible component type, but covers the anomalous conditions that are susceptible to the failure mechanism *and* could credibly result in severe consequences. The natural tension between the need to minimize false positives as well as false negatives must be resolved in favor of minimizing false negatives.

The result of generalization is a well-scoped set of anomalous conditions that envelope the potential for risk-significant occurrences of the observed anomaly failure mechanism throughout the system and at different times (for the affected component type).

Generalization is important even for (perhaps especially for) anomalies with little or no direct safety impact. A failure that is benign when it occurs under one set of circumstances may not be benign under another set. Figure 3-5 illustrates this by showing a relatively benign anomaly (low potential for severe consequences) whose causal mechanism might well have led to severe consequences (high potential for severe consequences) were it to have occurred differently. Thus, each anomalous condition can be independently assessed in subsequent APA process steps, without particular regard for the specific anomaly that prompted the analysis to begin with.

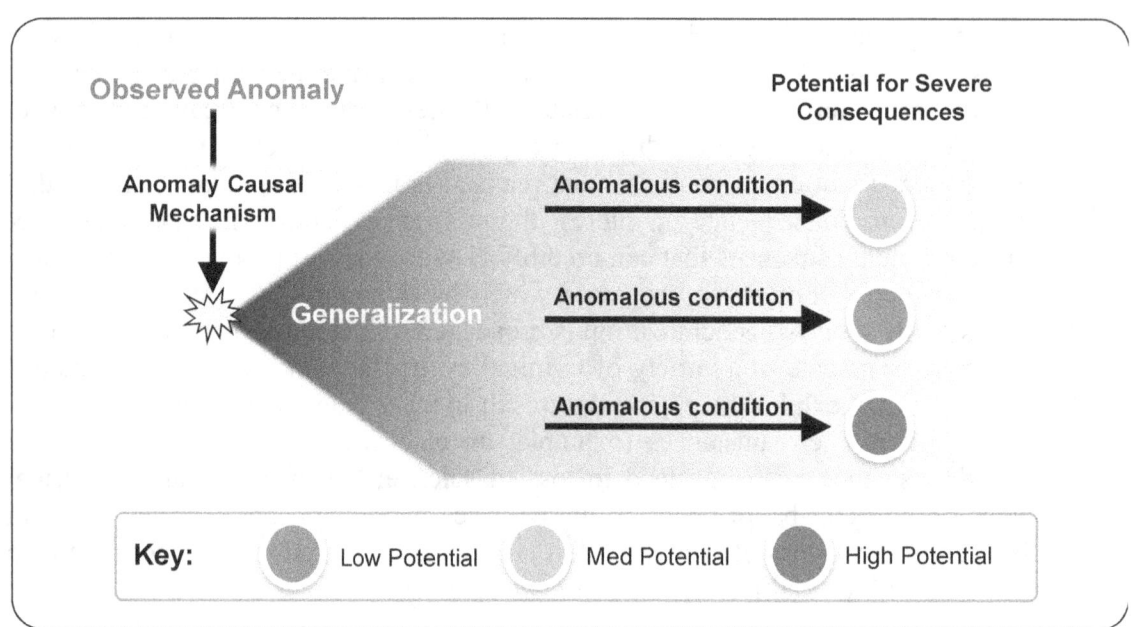

**Figure 3-5 Anomalous-Condition-Dependent Consequence Potential**

Although all anomalous conditions can be forwarded for further evaluation and grading, the fact is that time and resources can be better spent on focusing on those that have the highest potential of having severe consequences. Therefore it is recommended that once a sufficiently comprehensive set of anomalous conditions has been defined, the team reviews it for those bounding anomalous conditions it recommends be further evaluated. Since the point is ultimately to identify the risk significance of the failure mechanism, the crux of this exercise is to select the anomalous conditions that the team believes show the greatest potential for propagating to a severe consequence at this early phase in the process (see APPENDIX C for a DAnGERS screenshot which shows such a selection). Keep in mind that this process should not be bogged down by the exactness of the selections, so if there is any doubt which anomalous conditions could be more dire, then all those for which no distinction can be made in regards to potential severity should be retained.

### 3.2.6 The Generalization Team

As indicated in Table 2.1, Generalization is a team activity involving, at a minimum, precursor analysis experts to act as facilitator and scribe, and systems engineers with a comprehensive knowledge of system function, operation and system integration.

The activity is led by a facilitator, who challenges the team to consider the different circumstances in which the failure mechanism might occur. The facilitator balances comprehensiveness with time efficiency, making sure that the group considers a sufficient variety of circumstances, while also preventing the group from getting bogged down by over-generalizing the failure mechanism outside the bounds specified in Section 3.2.2. The skill and dogged focus of the precursor analysis expert in the role of facilitator is critical to successfully conducting generalization. The facilitator should be well versed in the nuances of the accident precursor analysis technical approach and somewhat familiar with the system to help draw out relevant insights from the subsystem experts. The facilitator should also play devil's advocate and encourage dissenting opinions to avoid a "group" think environment.

The role of scribe also plays a crucial support role in properly capturing the group's rationale at various points in the process in enough detail to make it comprehensible to someone reading the various justifications and descriptions in the future, but concise enough so that the process is not bogged down by endless documentation. The scribe should be familiar with the tool (e.g., DAnGERS) used to record the session and be able to navigate it as needed to support deliberations.

In application, the number of participants in a generalization session will vary. In general, proceedings will move along smoothly with a minimum of a facilitator with APA expertise, a system expert, a safety engineer and a scribe. Fewer than this minimum does not properly provide the quorum needed to effectively drive the process, and more (e.g., three system experts) can lead to extended deliberations, although the extended discussions could, if properly structured, lead to a better understanding of the potential effect of the failure mechanism. When generalizing anomalies from complex systems, it is often found that having multiple systems engineers present (each being expert in different subsystems) is beneficial to generalizing the causal mechanism across various subsystems.

***Space Shuttle Example of Generalization***

The Space Shuttle anomaly database contains an anomaly involving a sluggish helium reaction control system (RCS) fuel regulator. During generalization, the anomaly failure mechanism was identified as contamination, and generalized to the following anomalous conditions:

- RCS helium regulator fails closed due to contamination;
- Orbital Maneuvering System (OMS) helium regulator failure to open due to contamination;
- OMS helium regulator failure to close due to contamination.

The figure below illustrates the characteristics of the anomaly generalization process with a sample set of anomalous conditions. Each of these conditions was qualitatively evaluated to determine the point in the mission profile where its occurrence had the greatest potential to produce critical results. Subsequent analysis determined that the anomalous conditions generalized to the OMS were more risk significant than those that were applied to the RCS.

**Anomaly Report:**

Sluggish RCS He Regulator. PRI (PR301) Sys designator regulator (B-leg) was sluggish, pressure dropped below 200 PSIG during flow (suspect vent sensing line is clogged/blocked).---Under Review--- Problem analysis determined contamination from fuel vapor (due to improper seal) as most likely cause.

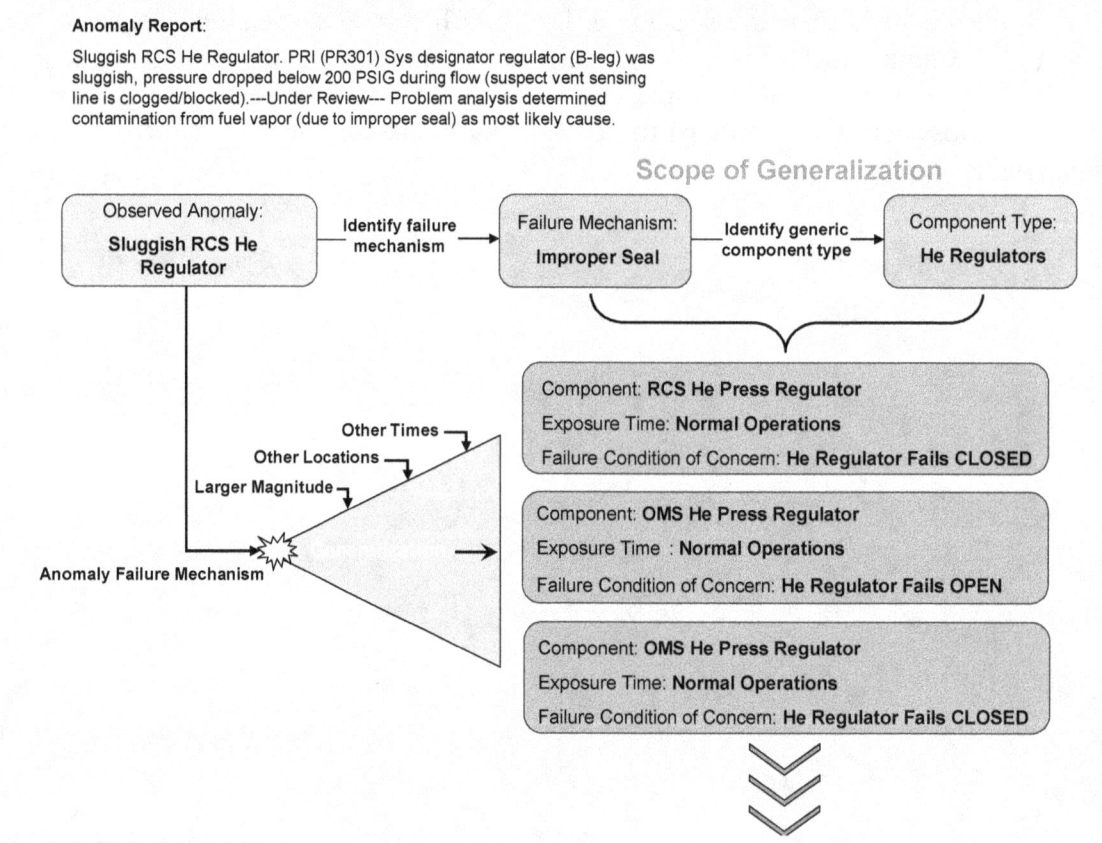

## 3.3 Evidence Gathering

The output of generalization is a set of anomalous conditions deemed worthy of further evaluation and grading. Subsequent grading of these anomalous conditions involves an evaluation of their potential to produce severe consequences. This potential depends on the physical locations and functional roles of the affected components.

Prior to grading an anomalous condition, data on each of the affected subsystems is collected, as well as sufficient system-level data to evaluate the potential for each failure condition of concern to occur, given the presence of the failure mechanism, and the potential for the failure condition of concern to propagate to severe consequences. The purpose of gathering evidence at this step is to root the decisions made during grading in some form of concrete and traceable data. Ideally this data will be in the form of engineering drawings, test results, physical analyses, etc, and will provide a more definitive basis than a speculative one by which to make judgments.

The task of evidence gathering serves as a break between the processes of Generalization and Grading (see Figure 2-5). It is an activity that can be conducted on an individual level, or split up and assigned to individuals of an evidence gathering team. As illustrated in Table 2.1, in order for the evidence gathered to provide useful information to the grading process the personnel must require competency in data mining and data interpretation and at-least moderate technical knowledge in the relevant engineering field or subsystem. In application to STS and ISS programs, it has been found that the most useful information is gathered by personnel who are knowledgeable in both system documentation and system-level effects of failure.

Since there is no distinction between anomalies and postulated anomalous conditions at this stage, the individuals or teams gathering the data treat each record equally. This is important since the data collectors may be swayed by the rigor they put into searching for relevant information based on whether they perceive the anomalous condition to be a "real anomaly". The data collected in this step is in addition to the anomaly-related data that has already been collected and used for purposes such as identifying the anomaly failure mechanism.

Because of the important role that evidence plays within the APA process, the data that is collected must be judged by the grading team as relevant to the analysis of the failure mechanism under consideration. It is not enough to associate data with a failure condition of concern if it does not bear in some direct way on the likelihood of the failure due to the mechanism, the propagation of that failure through the system, or the realization of severe consequences. Table 3.1 lists some of the data types that may be useful for understanding the potential system-level effects of failure, as well as the minimum criteria for inclusion in the collection.

The tool for conducting APA, DAnGERS, provides reporting features such as an 'Evidence Gathering Report' to support data mining activities. This report lists all anomalous conditions recommended for grading that are awaiting evidence; each anomalous condition is presented clearly and characterized by its relevant properties (i.e. failure mechanism, susceptible component, time frame, failure condition of concern and severe consequence). It is provided to data mining personnel to help guide the process of evidence collection, to provide an accident sequence structure on which to build evidence around, and to organize the collected documentation into legible 'evidence cases' for each anomalous condition. For further details on the evidence gathering report and other DAnGERS reports, see Appendix D.

The data collected should be organized in a manner such that it can be easily accessed during the grading session in which it is used. The set of evidentiary data may be delivered to the grading team in hard copy or digital format, as is available for each piece of evidence. To facilitate its ease of use, the evidence should be cataloged with respect to the relevant anomalous conditions, and listing the evidence title, document number, and the type of evidence. If the specific piece of data is available in an online database, a link may also be provided. Figure 3-6 shows an example representation of the organized output of the Evidence Gathering step.

**Table 3.1 Common Data Types and Minimum Criteria for Inclusion of Data**

| Data Type | Minimum Criteria for Inclusion |
|---|---|
| Failure Mode and Effects Analysis (FMEA) | Addresses failure mode involved in the accident progression |
| System Schematic | Shows components and interrelations involved in the accident progression |
| Design Drawing | Shows components and interrelations involved in the accident progression. Shows physical properties assumed in phenomenological judgments. |
| Fault Tree Model | Basic events include failure of interest and credited control failures under same or substantially similar system configuration and phenomenological conditions. |
| Operating Procedures | Events trigger procedures via established protocols. Procedural actions impact accident progression. |
| Reliability Analysis | Addresses same or substantially similar system under same or substantially similar conditions. |
| Test Procedure | Test protocols either establish failure thresholds or span credible conditions. |
| Physics Model | Addresses same or substantially similar system under same or substantially similar conditions. |
| Post-Flight Inspection Report | Addresses failure mechanism or failure mode involved in the accident progression. |
| Probabilistic Risk Assessment (PRA) | Events include failure of interest and credited control failures under same or substantially similar system configuration and phenomenological conditions. |
| Root Cause Analysis | Addresses failure mode involved in the accident progression. |
| Failure Observed in Operation | Addresses failure mode involved in the accident progression. |
| Test Data | Data either establishes failure thresholds or span credible conditions. |

**Evidence Summary:** Anomalous Condition ID#: 2 - 3

| Evidence Title | Document # | Evidence Type | Link |
|---|---|---|---|
| Impact Testing of HRSI | T0089-004 | Test Report | http:/nasa.gov/testreport |
| Specification for Installation of Orbiter Landing Gear Doors | KSC - 00569 | Operating Procedure | http:/boeing.com/flightprocedures |
| Shuttle PRA ver. 2.0 | NASA - 853484 | Probabilistic Risk Assessment | http:/nasa.gov/SPRA 2.1 |
| MLGD Tile Damage | MER-26 (STS-98) | In-Flight Anomaly | http:/usa.com/MER-IFA |
| NLGD Design Drawing | SEF852934-001 | Design Drawing | http:/nasa.gov/drawings |
| Structural Specification of HRSI Tiles | LMM-836 | Structural Certification Specification | http:/lmm.com/TPSstructspec |

**Figure 3-6 Example Summary of Evidence Gathered**

## *3.4 Grading of Anomalous Conditions*

The handoff from the initial generalization session and the subsequent evidence gathering activity is a set of anomalous conditions which have been produced from identified anomaly failure mechanisms and recommended for grading by the generalization team. Once sufficient evidence has been gathered, a multidisciplinary team will evaluate each anomalous condition for its risk implications based on the collection of applicable data. The following section describes in detail the methodology behind this process, known as *Grading*.

> *Grading*
>
> Grading is the portion of the APA process that evaluates anomalous conditions and recommends them for either *risk modeling*, *observation and trending*, or *no further action*, based on their assessed potential to lead to severe consequences. The process involves team evaluation of that potential in an anomalous condition in the context of its potential to produce an identified failure condition of concern and subsequent severe consequences.

For each anomalous condition that is evaluated in the grading process, a risk metric referred to as the *Potential Problem Index (PPI)* is assigned via a deliberative evaluation process, as discussed below. The PPI is formulated so that anomalous conditions with a high PPI warrant further investigation of the causal failure mechanism and its effect on system risk; those with a moderate PPI warrant a recommendation for continued monitoring and trending of related anomalies that share the underlying failure mechanism; and those with a low PPI are graded out of the process as being of low significance to the risk metrics being addressed.

Once an anomaly's failure mechanism has been generalized into a set of anomalous conditions, and the requisite evidence has been collected, the grading process can be conducted. Each anomalous condition is analyzed through the process of grading on an individual basis; the evidence that is applicable to the anomalous condition is considered and informs the assignment of a Failure Condition Index (FCI) and Conditional Consequence Index (CCI).

> *Failure Condition Index (FCI)*
>
> A qualitative measure of the likelihood for the specified failure condition of concern, given an occurrence of the anomaly failure mechanism on the specified component and at the specified time.
>
> *Conditional Consequence Index (CCI)*
>
> A qualitative measure of the likelihood for severe consequences, given an occurrence of the specified failure condition of concern.

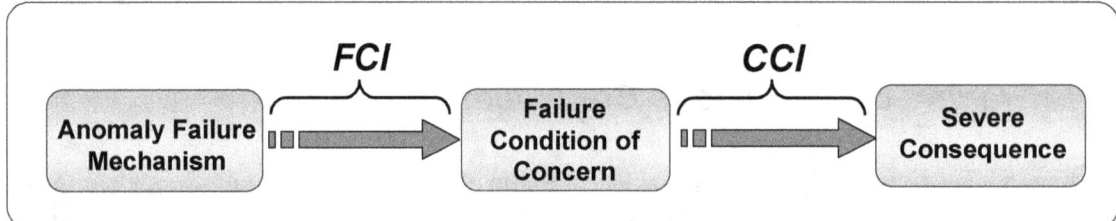

Figure 3-7 FCI and CCI in the accident sequence

It is important to keep a scenario-based perspective in mind when assigning FCI and CCI values to make sure that they are consistent with the definition of the failure condition of concern, i.e., the failure condition of concern upon which the CCI is conditioned should be the same as that used to assign the FCI. The assignment of FCI and CCI values is discussed in subsequent subsections. The FCI and the CCI are combined, along with the evidence caliber discussed in Section 3.4.3, to generate the PPI.

*Potential Problem Index (PPI)*

The PPI is the metric used in the APA process to grade an anomalous condition for either *risk modeling*, *observation and trending*, or *no further analysis*. The PPI is built up from three underlying metrics: the Failure Condition Index (FCI), the Conditional Consequence Index (CCI), and the Evidence Caliber (EC). The PPI takes into account:

- The potential for the anomalous condition to propagate from the anomaly failure mechanism to the failure condition of concern
- The potential for the failure condition of concern to propagate to severe consequences
- The caliber of the evidence used to support the above two potentials

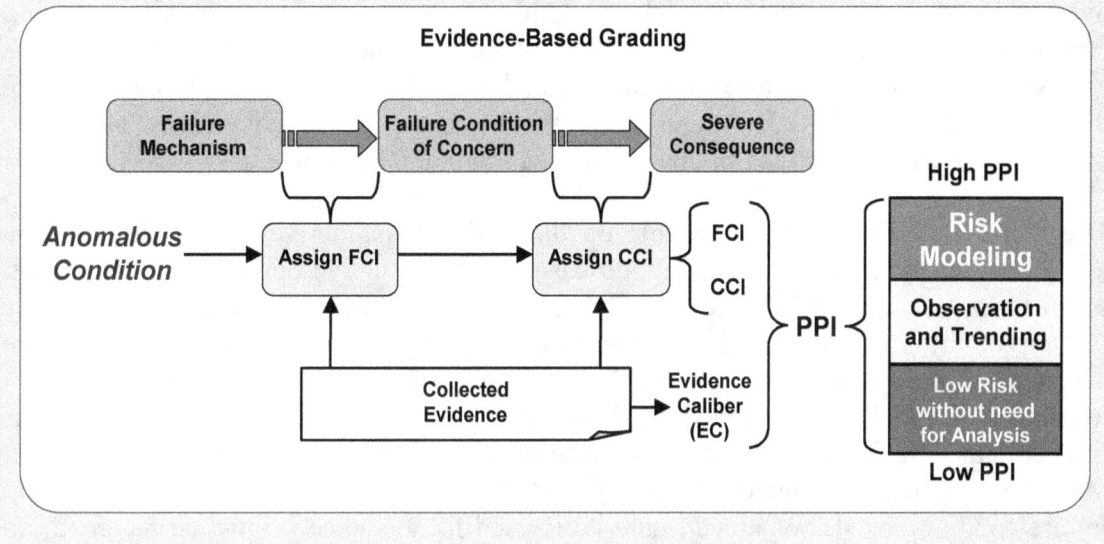

## 3.4.1 Assigning the Failure Condition Index

The FCI is assessed as the likelihood that an anomalous instance of the anomaly failure mechanism will propagate to the point where the threshold for the failure condition of concern is exceeded. This includes consideration of stressors that could exacerbate the failure mechanism and its propagation. It is important to remember that the context of the assessment is the postulated occurrence of the anomaly failure mechanism in a more severe manner and *not* the actual anomaly that was initially observed. Assessing the FCI of an anomalous condition involves three considerations:

1. The possible magnitudes of the anomaly failure mechanism

2. The threshold for the failure condition of concern

3. The ways in which the anomaly failure mechanism can propagate to produce conditions that exceed the threshold for the failure condition of concern

With regard to the threshold for the failure condition of concern, the range of conditions that the failure mechanism can propagate to needs to be assessed relative to the failure threshold.

The FCI assessment is *conditional* on the existence of the anomaly failure mechanism in the specified component. That is to say, the FCI does not take into account the likelihood for the failure mechanism to occur, only the likelihood that it can progress to the failure condition of concern. This conditional approach is prescribed for a number of reasons. First and foremost, it is conservative. At this stage in the process the concern is making coarse assessments of anomalous conditions to justify which need further detailed analysis and which do not. False positives are preferable to false negatives and possibly overlooking an underappreciated risk. Another reason for the conditional approach is economy. The grading part of the APA process is intended to be a quick triage assessment for identifying areas for further analysis. By taking a conditional approach, the assessment can be performed without considering the likelihood of the failure mechanism existing, and thus can be performed more quickly and easily.

The FCI is intended to be assignable on the basis of qualitative information, without explicit calculation. The FCI is assigned based on a qualitative descriptor, from a given set of terms, which best indicates the potential for the failure mechanism to produce the failure condition of concern. Each term indicates a particular susceptibility for failure, accounting for the severity spectrum of the mechanism and its effect at impacted locations. The FCI should be informed by existing data and supporting evidence. This might include fault size data for past occurrences, assessments of the bounding fault magnitude, damage thresholds for safety systems, or critical fault sizes above which a failure condition exists. Whatever material is used for evidence it must be recorded, not only for basic configuration management but also because the caliber of the evidence plays a role in the ultimate grading that the anomalous condition receives. Each qualitative descriptor is associated with a numeric FCI value. Table 3.2 presents the FCI assignment table used in ISS application of APA, which (along with the corresponding CCI table discussed in the next section) has been strongly influenced by the ISS Program Risk Scorecard [16], along with work in [8] and [9].

**Table 3.2 FCI Assignment Table**

| Failure Condition Index (FCI) | | |
|---|---|---|
| Descriptor | Index | Description of Basis |
| Very Likely | 6 | The failure condition of concern is expected to happen given the occurrence of the anomalous failure mechanism |
| Likely | 5 | It is likely that the anomalous failure mechanism could cause the failure condition of concern |
| Possible | 4 | It is possible that the anomalous failure mechanism could cause the failure condition of concern |
| Unlikely | 3 | It is unlikely that the anomalous failure mechanism could cause the failure condition of concern |
| Highly Unlikely | 2 | It is highly unlikely that the anomalous failure mechanism could cause the failure condition of concern |
| Non-Credible | 1 | The failure condition of concern is not credible given the occurrence of the anomalous failure mechanism |

The assigned qualitative FCI descriptor must ultimately be based on evidence, leading to the question of the caliber of the evidence used in the assignment. Where the evidence is weak, uncertainty is potentially large, and the "true" significance of the anomalous condition as a potential precursor might be higher than its assessed value would suggest. To accommodate this uncertainty, grading depends not only on the assigned FCI and CCI values, but also on the type of data used to support the evaluation, and its applicability to the anomalous condition under consideration (screenshots of the DAnGERS tool and how it records evidentiary data are available in Appendix C).

As codified in the DAnGERS tool, each generic data type is given a pre-established *Baseline Data Type Caliber* as a measure of the maximum credit that can be taken for the data, based on how generically useful or relevant that data type is expected to be. This is relative to the need in APA to ground the conclusions of the grading team in objective evidence. For example, test data has the potential to be very relevant as evidence, as test data can correlate the severity of a failure mechanism to a variety of relevant parameters; conversely, existing reliability analyses may be less relevant as evidence, because they represent the current state of knowledge regarding the system, and may not capture unknown vulnerabilities to failure mechanisms or the conditions they may create. In general, appropriate values for the caliber for each specific type of data may vary from one program to another (as might the type of data itself), for heritage reasons or reasons of analysis methodologies. When establishing an APA process for use with a system, a list of available data types should be assembled, and baseline data type calibers should be assigned based on the judgment of the precursor facilitators and system or program experts.

> **International Space Station Example**
>
> The FCI and CCI baseline data types and respective data type caliber values in the table below were developed for application to the ISS Program based on experience with that program's data types and underlying methodologies. These values were incorporated into the ISS DAnGERS tool to support FCI and CCI evaluation.
>
> | Type of Data | Baseline Data Type Caliber |
> |---|---|
> | Test Data | 90% |
> | Operational Observation | 90% |
> | Root Cause Analysis | 75% |
> | Probabilistic Risk Assessment | 70% |
> | Physics Model | 60% |
> | Test Procedures | 50% |
> | Reliability Analysis | 50% |
> | Operating Procedures | 50% |
> | Fault Tree Model | 40% |
> | Design Drawing | 40% |
> | System Schematic | 30% |
> | Engineering Judgment | 25% |
> | Failure Modes and Effects Analysis (FMEA) | 20% |
> | Educated Postulation | 10% |

During evaluation of FCI, the evaluation team identifies the data upon which they are basing their FCI assignment, and for each piece of data they assign an applicability rating, indicating the degree to which the data addresses the scenario(s) under consideration. Strong objective evidence (i.e., evidence with a high baseline data type caliber) might not directly address the accident conditions under evaluation (e.g., the regime of high fault magnitude where the potential to realize the failure condition or concern is greatest). Or, even when conditions match the postulated accident conditions, the uncertainty might be too high to confidently support a low FCI value, for example when test data in the appropriate regime is sparse. The applicability of the data is assessed qualitatively, with each rank corresponding to a percentage. Five ranks are used: strongly applicable, mostly applicable, somewhat applicable, slightly applicable, and minimally applicable. These applicability assignments, along with similar assignments for the data supporting CCI evaluation, are ultimately used in conjunction with the FCI and CCI values to determine the PPI and the corresponding grading of the anomalous condition. This process is described in more detail in Section 3.4.3.

It is critically important that the evidence used to support FCI and CCI assessments be reviewed and understood. Often, failures occur due to "unknown unknowns". That is, an attempt has been made to study the system and understand all of the operational uncertainties and vulnerabilities, but the understanding may be incomplete or incorrect. Because of this, it is the responsibility of the grading team to review the evidence used with a critical eye, questioning to make sure that the evidence presents a complete picture and that nothing was missed in the original analysis. If there is doubt within the group that the complete picture is understood, that should weigh in to the grading assessment for both FCI and CCI, again remembering to be conservative (false positives are preferable to false negatives).

## 3.4.2 Assigning the Conditional Consequence Index

The Conditional Consequence Index indicates the potential that the failure condition of concern will result in severe consequences. This is a function of the degree to which safeguards or barriers are present to mitigate the consequences given the failure condition of concern. If no such safeguards or barriers are present then the CCI should indicate that the severe consequence is expected. If they are present, then assigning a CCI value involves assessing the potential for these safeguards and barriers failing.

Safeguards and barriers that could mitigate the consequences can involve:

- Additional safety systems (levels of redundancy)

- Recovery actions

- Physical separation from an accident source

- Shields preventing penetration

- Time constraints on the propagation process

Assessment of the potential of particular safeguards and barriers failing is application dependent. Table 3.3 gives a list of qualitative descriptors used to designate the potential and associates each with a CCI value. Like the FCI, CCI assignment should be dependent on available data and analysis, which should be recorded as part of the assessment process.

**Table 3.3 CCI Assignment Table**

| Conditional Consequence Index (CCI) | | |
|---|---|---|
| Descriptor | Index | Description of Basis |
| Very Likely | 6 | Give the failure condition of concern the severe consequences would be expected |
| Likely | 5 | It is likely that severe consequences would result given the failure condition of concern |
| Possible | 4 | It is possible that severe consequences would result given the failure condition of concern |
| Unlikely | 3 | It is unlikely that severe consequences would result given the failure condition of concern |
| Highly Unlikely | 2 | It is highly unlikely that severe consequences would result given the failure condition of concern |
| Non-Credible | 1 | The failure condition of concern could not credibly give rise to severe consequence |

Like FCI, the assigned qualitative CCI descriptor must ultimately be based on evidence. Hence, the same process for assessing data applicability is used for CCI evaluation that was used for FCI evaluation. The types of data that are generically relevant to CCI assignment are identical to those relevant to FCI assignment, as are their baseline data type calibers, as shown in the previous example.

### 3.4.3 Grading Results

In the grading step, a numerical risk score is computed based on consequence indices and data calibers and applicability levels determined in the previous steps, using a scoring rule that intends to capture the impact of the consequence indices and evidence caliber on the perception of risk.

The risk score is primarily a function of the FCI and CCI scores. Under the assumption that the steps of the FCI and CCI represent order of magnitude changes in the respective potential levels, the FCI and CCI scores are summed to obtain an overall problem potential measure.

Additionally the scoring rule applies an upward adjustment when there is limited evidence available, to reflect the level of uncertainty as part of the Problem Potential Index (PPI): if the data type caliber of all the data used to evaluate FCI and CCI were 100%, and all the data were strongly applicable to their respective evaluations, then the PPI could be determined directly from FCI and CCI, and in fact would simply be the sum of FCI and CCI. For decreasing evidence calibers, progressively larger positive adjustments are made, resulting in correspondingly higher PPI values for the same FCI and CCI values.

The evidence caliber itself is a function of the baseline data type calibers and assigned applicability of the supporting data used by the evaluation team when assigning FCI and CCI. For each piece of data, its applicability (expressed as a percentage) is multiplied by the baseline data type caliber to produce a *Data Caliber*, DC.

$$DC = datatypecaliber * applicability$$

Then, the DC metrics are aggregated together to produce an overall *Evidence Caliber*, EC, which ranges between 0 (no evidence) and 1 (fully informative evidence). This is done in such a manner that incremental pieces of evidence will drive the EC towards 1. However, as EC gets closer to 1 (representing a fully informed state), additional information will have a diminished impact:

$$EC_{FCI} = 1 - [(1-DC_1)*(1-DC_2)*...*(1-DC_M)]$$

**Equation 1**

$$EC_{CCI} = 1 - [(1 - DC_{M+1}) * (1 - DC_{M+2}) * ... * (1 - DC_N)]$$

**Equation 2**

$$EC_{PPI} = EC_{FCI} * EC_{CCI}$$

**Equation 3**

In the above equations, M pieces of data support FCI assignment (Equation 1) and $N - M$ pieces support CCI assignment (Equation 2).

This simplified evidence caliber rule simply treats each piece of evidence as additive, and does not account for conflicting information.

Equation 1 through Equation 3 define the data caliber aggregation within a single index ($EC_{FCI}$ and $EC_{CCI}$, respectively), and for the two indices combined ($EC_{PPI}$). Aggregation within an index is treated as a co-product, under the principle that multiple sources of evidence contribute to an increasing confidence in the assigned value. Aggregation across the indices is treated as a product, under the principle that each index is separately supported by its evidence, so that the evidentiary support for PPI as a whole is no stronger than its weakest link. If $EC_{PPI}$ is low then one can conclude that there is some uncertainty in PPI, and therefore the grading should be correspondingly conservative in this case. The equation for PPI is given by the following:

**Equation 4**

where $c$ is a constant that controls the impact of the data caliber on the risk score. During the precursor pilot exercises, a value of $c = 1.1$ was adopted, causing the difference between no evidence and fully informative evidence to be approximately one step on the potential scale.

The PPI is the grading metric for each anomalous condition which is the basis for the recommended further action as shown in Figure 3-8. A high PPI is used to denote an anomalous condition with a high level of perceived risk, and correspondingly a low PPI denotes a low level of perceived risk. The grading result specifies:

- Risk modeling in the case of a high PPI;

- Observation and trending, in the case of a moderate PPI; and

- No further analysis, in the case of a low PPI.

For high values of evidence caliber, PPI closely matches the sum of FCI and CCI. For low values of evidence caliber, PPI is higher than the sum of FCI and CCI, indicating an upward adjustment to account for the higher uncertainty inherent in the FCI and CCI assignments. To account for the differences between specific programs there are no absolute thresholds set for the division between each further action. For example, an anomalous condition which receives a PPI grade of 8.2 might be considered lower risk by some programs, and yet high risk by others. Instead the thresholds should be set on program-by-program basis, tailoring to the specific needs of the program. One method for doing this is by applying APA to known issues as benchmarks, and using the expected outcome as a way by which to set threshold levels.

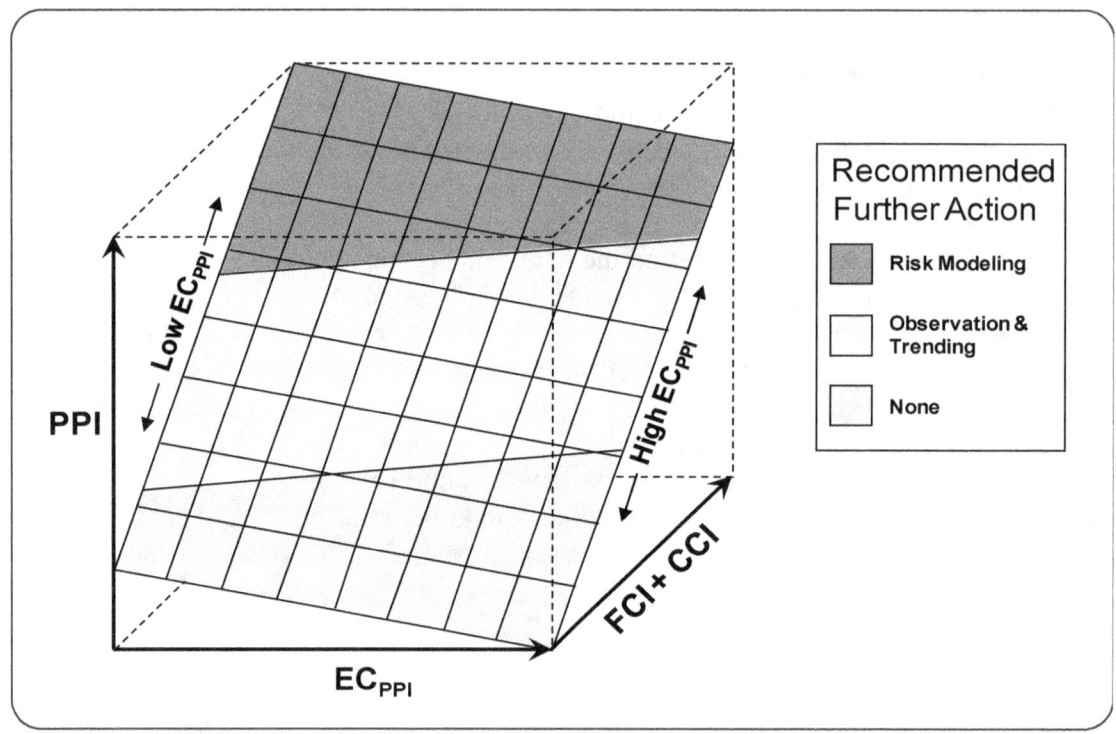

Figure 3-8 Notional Graph of Recommended Further Action as a Function on FCI, CCI, and $EC_{PPI}$

The lowest possible graded result simply indicates that the anomalous condition is not risk significant. The anomalous condition was recommended for grading; however, upon evaluation, the assessed PPI indicates that the anomalous condition is within the safety envelope of the system and does not warrant further analysis.

If the PPI of an anomalous condition is judged not to be high enough to warrant explicit risk modeling, but also is non-negligible, then an intermediate grading recommends observation and trending of the failure mechanism to ensure (1) that the full history of the failure mechanism is brought to bear on the characterization of the spectrum of fault magnitudes; (2) that parameters correlating to anomaly occurrence and fault magnitude are identified; and (3) that adverse trends in frequency or fault magnitude are identified before they penetrate the safety envelope of the system. Once the finding is made, if the recurring problem analysis determines that the anomaly is not part of a recurring set of events, it is possible that the failure mechanism is deemed to be a low risk, or in other words "not risk significant." However, if there is anomaly trend that raises questions then additional modeling may be necessary to understand the trending pattern to ensure that the failure mechanism poses no significant risk to the system, or to redesign the system to preclude recurrence.

The highest grading possible, risk modeling, is prescribed for those circumstances where the potential for severe consequences is too high to simply continue observation and trending. The rationale is that a high PPI for a failure mechanism that has actually occurred warrants an explicit investigation into its risk implications. Risk modeling is discussed further in Section 3.6.

### 3.4.4 The Grading Team

As indicated, grading is a team activity involving, at a minimum, precursor analysis experts to act as facilitator and scribe, systems engineers with a comprehensive knowledge of system function, and safety personnel who understand the subsystem and system-level effects of failures. The environment is that of structured brainstorming, similar to that used to conduct FMEAs and Hazard and Operability studies (HAZOPs).

The activity is led by a facilitator, who moderates the evaluation of each anomalous condition, making sure that all voices are heard, and who fosters consensus on FCI and CCI designation in terms of reasonable, conservative values given the spectrum of opinion and variety of data that are present. The facilitator assures that key assumptions and results upon which the team's reasoning depends are supported by documented evidence. As with generalization, the facilitator balances comprehensiveness with time efficiency, in this case making sure that the group evaluates each anomalous condition to an appropriate degree, while also preventing the group from getting bogged down or sidetracked, e.g., due to differences of opinion or speculation in the face of limited data. As the facilitator steps through each anomalous condition, systems experts and safety engineers evaluate the potential for progression from failure mechanism to failure condition of concern, and from failure condition of concern to severe consequences.

## *3.5 Observation & Trending*

Anomalous conditions that are perceived to represent a moderate potential for producing severe consequences are recommended for continued Observation & Trending. Observation & Trending is the process of tracking and analyzing related anomalies over time, in order to identify trends in the recurrence of anomaly failure mechanisms, before they manifest themselves as more severe problems. Trending anomaly data can include:

- Trends in the rate of anomaly recurrence (frequency)

- Trends in fault magnitude as a result of anomaly failure mechanisms

- Trends in the Potential Problem Index (PPI trending)

Trending anomaly occurrences can help shed light on the time-dependent behavior of systems and thus refine the understanding of anomaly failure mechanisms at work. Trends in the data can be both 'positive' and 'negative'; it can tell us that a causal mechanism is recurring with an increasing frequency over time, which could imply that some corrective action for the component type experiencing the mechanism may be warranted. Conversely an anomaly that is recurring with a decreasing frequency over time may tell us that a correction action against the mechanism is performing correctly, or simply that system maturity or operating experience is precluding future instantiations of the mechanism. Either way, it is not the purview of APA to enforce actions based on interpretations of an anomaly trend, but rather to provide data that can help a program to perform trending in a meaningful way.

The APA process specifies that trending be performed for all anomalies graded for either Observation & Trending *or* Risk Modeling. This is because anomalous conditions that are regarded as being most critical should be given the full spectrum of analysis available, to allow decision makers to make the best-informed judgments regarding corrective/preventative actions.

As detailed in Section 3.2.3, a generalization group gathers together related anomalies, the anomalies in each generalization group represent suitable populations for the purpose of trending. Trending analysis requires a number of data-points (anomaly recurrences) in order to produce a trend, thus each generalization group containing multiple anomalies defines the foundation for a trending population. Figure 3-9 shows a notional collection of generalization groups (expanding on Figure 3-3) and their respective trending populations.

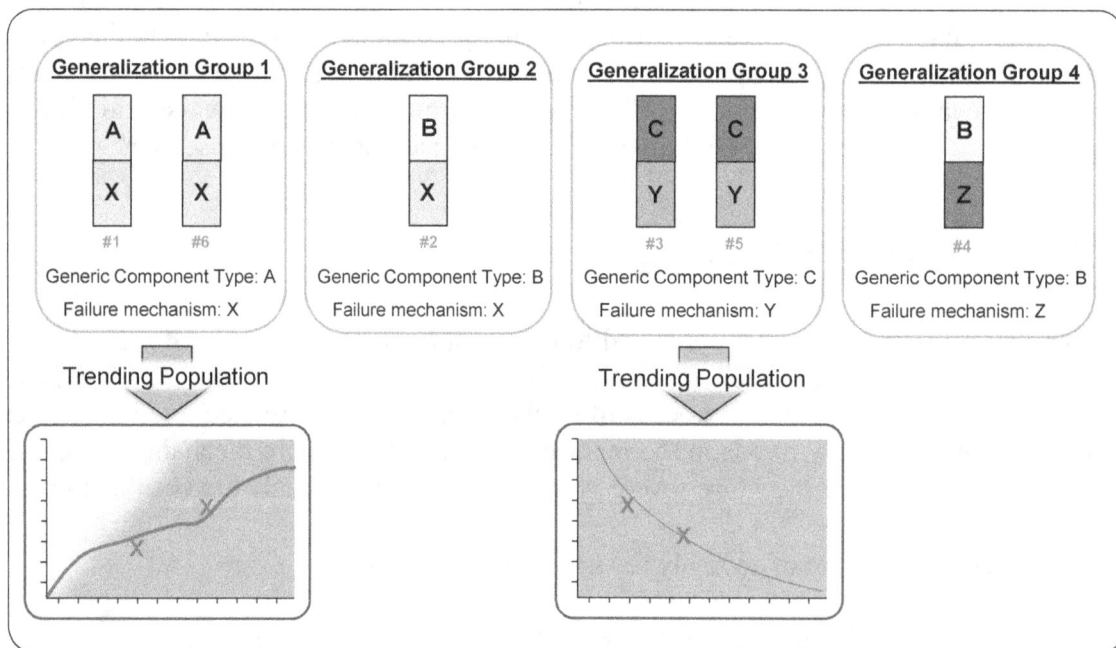

**Figure 3-9 Generalization Groups and Trending Analysis**

Whenever an anomalous condition is recommended for Trending, it is actually the observed anomalies within the same Generalization Group that are recommended for trending. Trending can only be performed on observed anomalies that actually occur within the system and NOT on any of the postulated anomalous conditions produced during the APA process of generalization. Every anomalous condition within a generalization group is fundamentally related by the anomaly failure mechanism acting on the generic component type; therefore any recurrences of anomalies within the group tells us something about the potential for the anomaly failure mechanism to recur within the generic component type. Thus, anomaly trends can be utilized to inform conclusions on the anomalous conditions within the same generalization group. Figure 3-10 illustrates this concept.

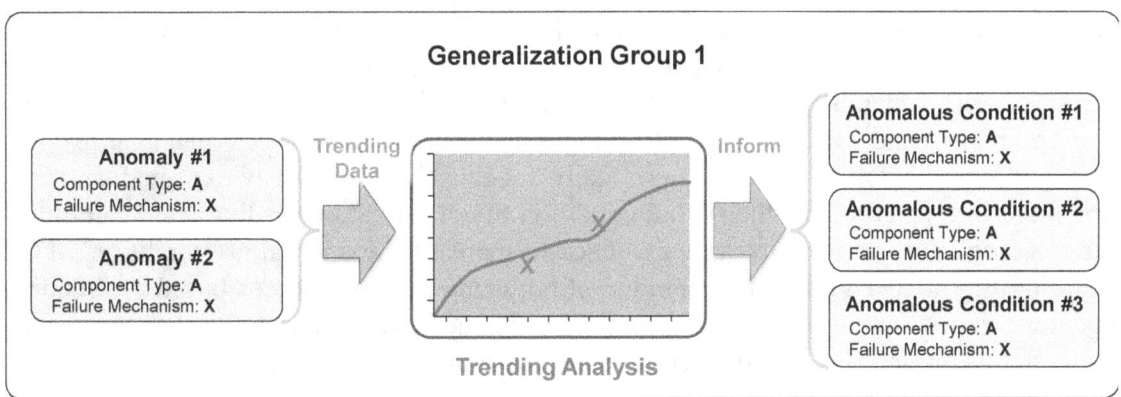

**Figure 3-10 Trending Informs the Group**

The extent to which trending can be performed depends largely on the availability of recurrent anomalies within a group. For example, if an anomalous condition is recommended for Observation & Trending or Risk Modeling and it is found that there are none or very few instances of recurring anomalies within the group, the population of data points in a trend may be insufficient to reliably infer any conclusions. Under such circumstances the trending activity may be limited to data gathering pending the generation of sufficient information to produce such a trending analysis.

> *Trending and the Generalization Group*
>
> The observed anomalies within a generalization group form the population for trending anomaly fault magnitude and frequency of recurrence. Trending analyses can be utilized to inform conclusions for all anomalous conditions within the same generalization group.

Trending the recurrences of related observed anomalies can be conducted using various analytical techniques. If a trending process already exists within the program then it should be utilized to perform the trending analysis. If such a trending activity is not established, there is some trending advice detailed in Appendix E.

### 3.5.1 Using Trending Data as Evidence

Once an anomalous condition goes through the APA process steps of Generalization, Evidence Gathering, and Grading it is recommended for one of three further actions. If that action is Risk Modeling then sufficient resources are brought to bear in order to characterize the overall risk significance of the underlying failure mechanism. If an anomalous condition is recommended for Observation & Trending, there lies the potential that over time, subsequent trending analysis will signify that it represents a higher risk to the system (higher than 'moderate'), and should warrant scenario-based risk modeling. APA supports a technique that enables adjustments to be made to such anomalous conditions based on trending results, in order to escalate its recommended action to the appropriate level. This technique is essentially composed of a feedback loop within the APA process.

As detailed in Section 3.4 (Grading Anomalous Conditions) the assignment of FCI and CCI is rooted in supporting evidence. If a trending analysis has been conducted (or updated) for a generalization group, it can now be utilized as evidence for assigning an appropriate FCI and CCI. For example, a trending analysis may indicate that a large crack in a fluid line (fault magnitude) is more likely to occur than previously expected, and thus it is more likely to produce the failure condition of concern. The grading team can utilize the trending analysis as supporting evidence to increase the anomalous condition's FCI, resulting in a higher PPI. This technique of adjusting grading indices based on trending updates ensures that the APA recommended further actions are kept up-to date and reflect the system's current risk as indicated by observed failure mechanisms.

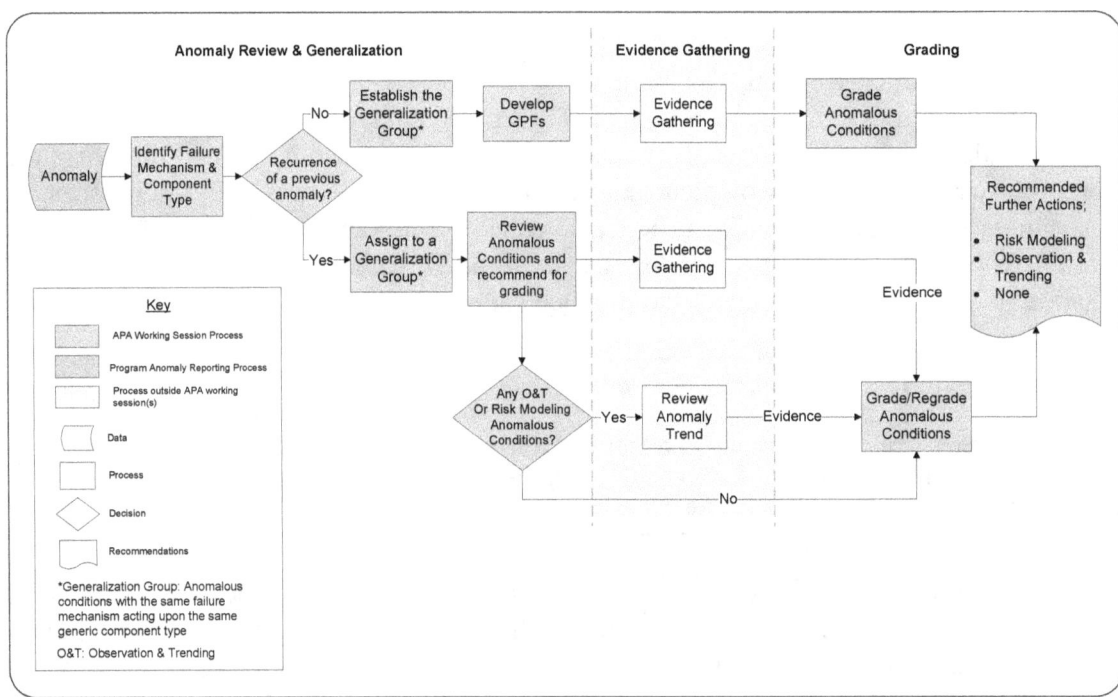

**Figure 3-11 Trending Data Feedback Process**

### 3.5.2 Trending Parameters

Trending can be carried out on a variety of anomaly parameters that are fundamental to the concept of recurrence; fault magnitude, PPI (which represents the perceived risk, in the context of APA), and frequency. It is useful to consider these parameters individually; as one parameter may signify a converse trend to any other given parameter. Any conclusions inferred from trending analyses must be based on the merits of the individual case (for example a decrease in anomaly frequency may actually signify an increase in system risk if the trend in fault magnitude is seen to be increasing to significant levels). Trending multiple parameters for a given group of related anomalies helps to provide a comprehensive set of data in order for informed decisions to be made regarding corrective action on recurring failure mechanisms.

### 3.5.2.1 Fault Magnitude Trending

One of the main objectives of the APA process is to identify anomalies which occur with benign magnitudes but might recur with a greater magnitude; representing a potentially greater threat to system safety. Trending the magnitude of an anomaly provides an informed understanding of this potential. It is important that trends in anomaly magnitude are interpreted on the specific merits of the scenario in question; trends (or lack of trends) can portend various characteristics of system operation. Trends that show decreasing magnitude in anomaly recurrence can provide evidential basis that a corrective action is successful; conversely the same decreasing trend may actually be in response to a reduced demand on the component and does not signify any reduction in system risk. A trend signifying increasing magnitude in anomaly recurrence may imply that the causal mechanism is growing with time (or system operation) or that barriers between the causal mechanism and component failure are deteriorating or ineffective.

In order to trend magnitude, the anomaly must entail a measurable physical parameter by which to apply a magnitude scale. Magnitude trending cannot be applied to binary types of anomaly e.g. a fuel-line valve fails open or closed, instead it would have to consist of a valve delay (measured in time) or mechanical sluggishness (speed of valve closure/opening). One of the initial tasks in fault magnitude trending involves establishing those measurable physical parameters, examples include:

- Crack size

- Leak rate (volume per unit time, volume per operation)

- Power supply over-current

- Valve open/close delay

Techniques for trending anomaly fault magnitude can vary based on both the system and the specific characteristics of the anomalies in question. It is not the purview of APA to prescribe trending techniques; it is the program's decision to implement techniques most appropriate to them. Figure 3-12 illustrates some sample trending techniques.

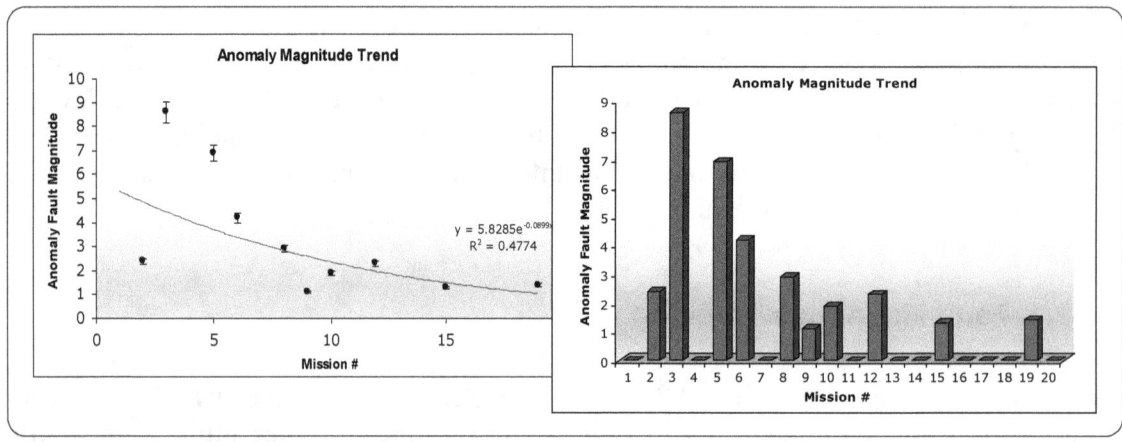

Figure 3-12 Sample Anomaly Fault Magnitude Trends

### 3.5.2.2 PPI Trending

PPI represents the perceived risk that an anomalous condition poses to the safety of the system (specifically the likelihood for the failure mechanism to propagate to the defined severe consequence). As discussed in Section 3.5.1, a trend in anomaly recurrences can be used as evidence in the Grading process to adjust FCI or CCI; once a PPI has been updated this way there is now a change in PPI. This adjustment of the PPI based on what the system is telling us today, is the mechanism that can create PPI trends. Each time an updated trending analysis is used as evidence to update the FCI or CCI, the PPI trend provides a record of the Grading team's interpretation of the potential risk that the trend and the anomalous condition presents.

Anomaly fault magnitude and anomaly frequency trends convey the pattern of failure mechanism recurrences on the specific component that has experienced the recurring observed anomalies (and possibly within a certain timeframe/mission phase). PPI trends do not show patterns of real-life anomaly recurrences, instead they illustrate patterns in the perceived risk of the postulated anomalous conditions created during generalization. Those anomalous conditions represent the possible instantiations where the failure mechanism could occur under different circumstances from the anomaly (different locations, magnitudes or times). Thus, PPI trends of anomalous conditions demonstrate how an anomaly trend is affecting system safety in other areas and other circumstances than would be communicated by anomaly trends alone.

The PPI trend is a useful method for communicating how recurring anomalies may impact risks at various locations and times throughout the system. Decision makers can utilize this piece of information to highlight areas of deteriorating safety and leverage it to garner the attention required to implement corrective action resources. A PPI trend does not infer any causation as to why the implied risk is increasing or decreasing; system experts will need to 'drill down' through the workings of the trend (anomaly frequency/magnitude, failure modes, causal mechanisms etc.) in order to identify the root cause of the trend.

Figure 3-13 demonstrates the PPI trend from a sample generalization group; recurring anomalies are shown along the bottom axis, occurring intermittently between grading sessions G1, G2, and G3. Anomalous Condition 1 is recommended for Observation & Trending in the first grading session and trending analysis is conducted for the recurring anomalies and is utilized in the second grading session as evidence for re-assessing the anomalous conditions' PPIs. The grading team's re-assessment of the PPI, in light of the new trending data, results in an increased perceived risk of propagating to severe consequences (increased PPI), and the anomalous condition is subsequently recommended for Risk Modeling.

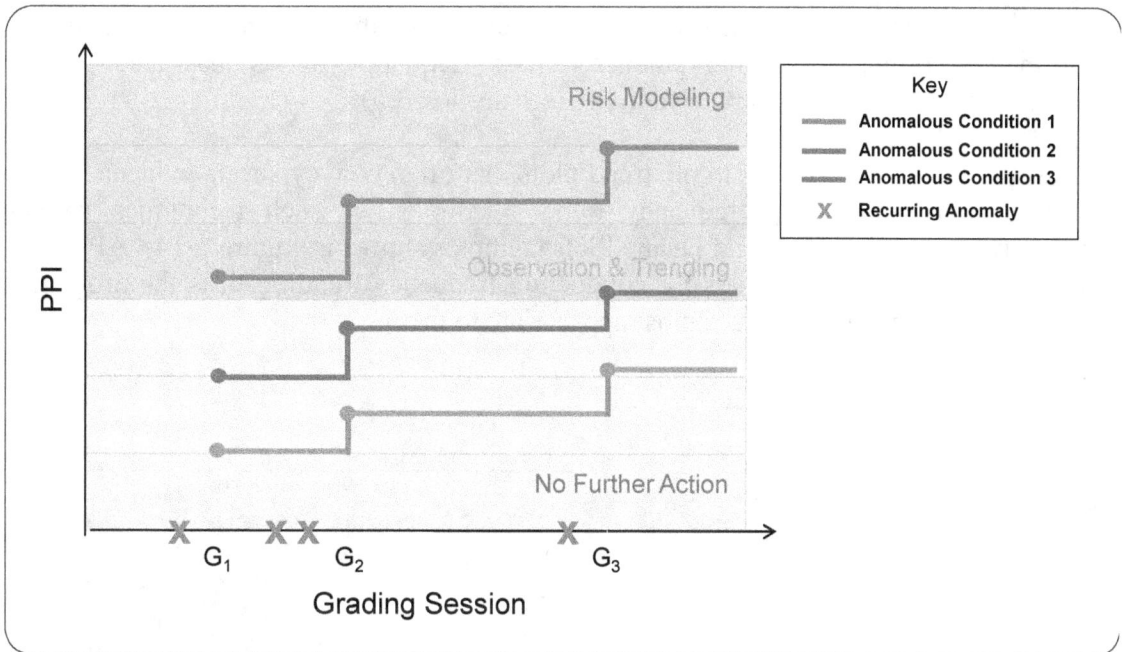

**Figure 3-13 PPI Trend of a Generalization Group**

### 3.5.2.3 Anomaly Frequency Trending

Trending anomaly frequency rates evaluates the pattern of recurrence of anomalies to determine the degree to which anomalies of that type are becoming more (or less) common. Frequency trends of anomaly recurrences can be indicative of many issues; they can indicate that a causal mechanism has not been fully investigated or that the causal mechanism has not been fully addressed or understood (perhaps a corrective action has been put in-place but the anomaly is still recurring). Without conducting trending analyses the occurrences of the failure mechanism may go without further investigation other than to 'safe' the system for immediate flight after each occurrence. Conducting a frequency trend of such anomalies with an appropriate trending basis may reveal that the causal mechanism is recurring with an increasing rate during operation and may warrant time and resources to properly identify and address the causal mechanism. Precluding future occurrences of recurring mechanisms not only reduces system risk but also minimizes resource expenditure for future 'safing' of the system for immediate flight. In accordance with the 'Graded' approach to systems engineering the cost of investigation/corrective action for recurring anomalies must be in accordance with the risk significance concluded from any trending analyses.

A frequency trend is not considered to be a relevant piece of evidence applicable to the APA grading process because the grading indices are conditional on the occurrence of the anomaly failure mechanism. This effectively assigns a probability of unity to anomaly occurrence, a conservative stance based on the fact that the failure mechanism is operative in the system, as evidenced by the anomaly. That is not to say that frequency trending is not useful in safety engineering and system reliability per se; it is just that frequency trends are not part of the feedback loop within APA. If a trending program which incorporates frequency trending already exists for the system, then it should be performed according to normal practice. Those frequency trends may also provide valuable input to the Risk Modeling activity (see section 3.6).

It is useful to construct a few simple trend plots that often convey strong evidence of the presence or absence of a frequency trend. Examples of such techniques include Cumulative Frequency Plots and Duane Plots (samples shown in Figure 3-14). APA does not prescribe specific techniques for conducting frequency trending; it is the program's decision to implement techniques most appropriate to them.

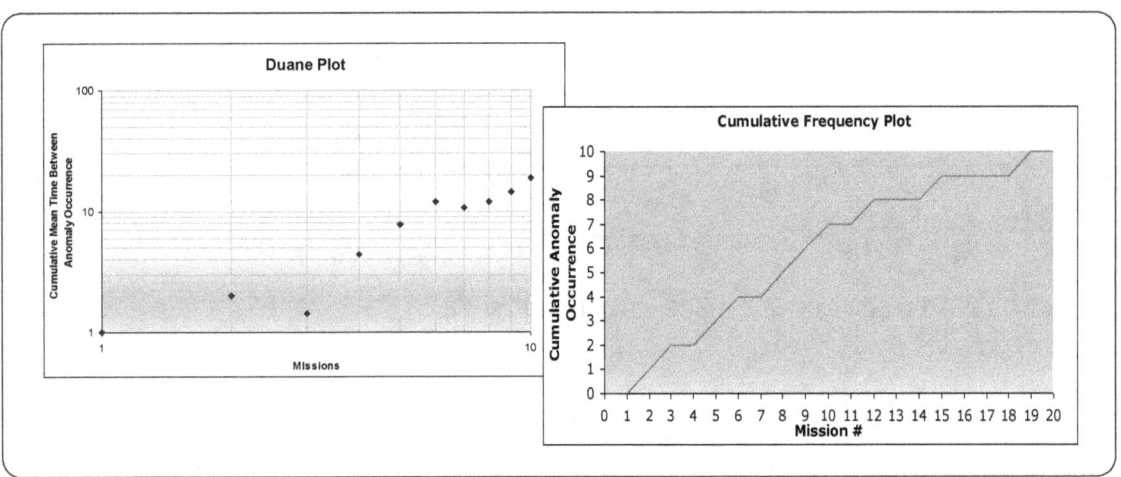

**Figure 3-14 Sample Duane & Cumulative Frequency Plots**

> ***Observation & Trending Summary***
> - Anomalous Conditions recommended for Observation & Trending <u>or</u> Risk Modeling are trended
> - Related anomalies are grouped by failure mechanisms acting upon component types (Generalization Groups)
> - If an anomalous condition is recommended for Observation & Trending or Risk Modeling, it is the observed anomalies within the generalization group that are trended
> - Trending Analysis includes frequency of occurrence as well as magnitude of occurrence
> - PPI is trended to communicate potential risk significance changes related to anomaly recurrences
>   - Decision makers can use this to drill down through the trending analysis to identify the trend driver(s) - In order to better focus corrective actions to have the greatest risk impact

## *3.6 Risk Modeling*

Risk modeling is indicated for anomalous conditions above the upper threshold for observation and trending. The rationale is that a high PPI grading of an anomalous condition warrants a deeper understanding of its risk implications. The grading step serves as a relatively rapid assessment of potential risk in order to determine which anomalous conditions need detailed attention, in the form of risk modeling. Unlike the grading process which was already completed, the risk modeling step is a *quantitative* assessment of the likelihood for severe consequence that is presented by the anomalous condition.

In accordance with the graded approach to system safety modeling specified in NASA/SP-2010-580 [6], the level of detail of the risk modeling should be commensurate with the magnitude of the risk. For example, if it is found that a risk model already exists for the anomalous condition, then the upshot of the risk modeling activity should be a potential change in the predicted frequency, reflecting the fact that elements of the affected scenarios have been observed, rather than a complete replacement of the existing risk model. To that effect, the risk modeling activity is meant to leverage existing analyses wherever possible. If a mature system risk model exists, it may be used to quantify the risk of the potential anomalous condition. If the existing model does not model the anomalous condition, then the condition should be modeled as it would integrate into the risk model.

If an existing system risk model does not exist, then the risk modeling step may be performed through any standard probabilistic modeling technique, chosen at the discretion of the risk modeler. These types of techniques may include, but are not limited to: Physical Simulation Modeling, Event Tree Modeling, Fault Tree Modeling, or Event Sequence Diagram Modeling.

### 3.6.1 Anomalous Condition Risk Importance

Risk modeling (in the context of APA) is performed to specifically measure the risk significance of the anomalous condition, and to do so in the context of a benchmark system risk ($R_o$). The APA process defines the Anomaly Condition Risk Importance (ACRI) measure to assist in gauging and prioritizing anomalous conditions in terms of risk. ACRI, the mathematical basis for which is discussed in detail in Appendix F, is the conditional risk that is directly attributable to a failure mechanism occurring outside nominal bounds (thereby creating an anomalous condition). The metric is not intended to provide a complete picture of the anomalous condition's risk, but merely a mean, comparative measure by which to take a top-level view of the risk. As its name implies, this risk metric is conditional, i.e. it presumes the anomalous condition exists. The ACRI is calculated as:

$$ACRI = [(Risk\,|\,AnomalousCondition) - (Risk\,|\,\neg AnomalousCondition)] / R_o$$

**Equation 5**

The second term in the equation for ACRI is needed to remove any risk contributions that are not directly attributable to the anomalous condition, such as that due to random failures of modeled safety systems. This also removes the unrelated contributions that are due to the choice of model scope. For example, a detailed model that includes all the components of a subsystem will typically show more risk than a subsystem model that is restricted to the components directly involved in the anomalous condition.

When a system risk model exists, the system risk $R_o$ can be taken from the model. In cases where a system risk model does not exist, $R_o$ must be obtained by other means in order to provide a normalization factor against which risk significance can be established. One possible basis for normalization, in the absence of a calculated risk, is the system risk requirement, which establishes a de facto acceptable risk. Note that assuming an $R_o$ of 0 (meaning the system has absolutely no risk) leads to an ACRI of infinity – this makes sense since a system with no risk should not be experiencing anomalies in the first place.

Keep in mind that it is imperative that the same benchmark be used in calculating the ACRI for all system anomalous conditions since this is actually a comparative measure that should be used for prioritizing where resources should be spent to maintain the benchmark level of risk. In this regard, an analogy is noted between the ACRI measure and the Risk Achievement Worth importance measure [17], which indicates features that are of most interest to maintain the present level of risk, and are thus of special interest in reliability assurance programs and inspection and enforcement activities.

If the benchmark is updated, it should be applied retroactively to all previous ACRI estimates to ensure a common yardstick. Note that since $R_o$ is a constant within the program/project that it is applied, pair wise comparisons in the form of ACRI ratios can also be used to assess the importance of anomalous conditions in relation to each other. In this case the actual value of $R_o$ becomes irrelevant since it factors out of the equation.

## 3.6.2 Risk Modeling Outcomes

The risk modeling activity entails assessing the likelihood of creating a severe consequence through probabilistic means, and modeling the relationships among sources of risk and their impacts. The model should aim to identify the sources and nature of risk in conjunction with the uncertainty associated with the anomalous condition. The goal of any risk modeling activity is to identify the parameter(s) or area(s) within the scope of the anomalous condition that is contributing the greatest level of system risk. By identifying the origins of risk, one can be better prepared to make decisions about how to mitigate it.

The outcomes of risk modeling activities should be reported to top-level management in an effort to make them aware of the failure mechanisms active in the system, and the operational vulnerabilities that can arise from them. By effectively communicating these risks, management can make risk-informed decisions about how the system should be operated and what design modifications must be made to ensure safe operation.

# 4 Documentation of Process Steps and Reporting Results

The process of analyzing observed anomalies to identify potential precursors involves numerous tasks: data mining, brainstorming activities, evidence gathering and working sessions. These tasks must be documented in order to provide a robust record of the analysis process and to communicate the results effectively so that informed decisions regarding the safe operation of the system can be made.

The purpose of reporting the risk results to top-level management is to:

- Make them aware of failure mechanisms in the system for which the risk may be underappreciated

- Highlight system vulnerabilities, thereby supporting risk-informed system modification

Management can decide, based on a risk-informed picture, how the system should be operated, and what design modifications must be made to ensure flight safety. The APA process does not prescribe definitive actions to be followed once an anomalous condition passes over a certain risk threshold, the process of correcting these potential problems and/or addressing the risk is specific to each program and its management.

Each program has its own style of risk management and reporting practices; however in practical applications of the process it has been found that there are generally 3 main steps to reporting APA results:

1. Declare the scope and timeframe of the anomalies reviewed (i.e. "All in-flight anomalies generated from Sept 1, 2009 to June, 2010")

2. Show how the anomalies were generalized and how the anomalous conditions were graded

3. Report the ACRI for all anomalous conditions graded for Risk Modeling

Details of the APA results can be presented in many formats, at the discretion of the management responsible for making APA-informed decisions, some examples of results used during pilot applications of the process are shown in Appendix G.

## 4.1 Reports from DAnGERS

At various points in the process, specific outputs in the form of reporting methods and formats have been defined to facilitate the dissemination of information in a standard and consistent fashion.

There are a variety of standardized forms and reports recommended as outputs from the APA process, and are automatically produced from the DAnGERS tool; example documents are shown in Appendix D:

**Document 1: Anomaly Review Report**
- A list of all the observed anomalies from the caseload
- Identifies which anomalies have been reviewed for generalization and which ones are awaiting review
- Communicates the level of progress (how many observed anomalies from the caseload have been reviewed) achieved by the Generalization team

**Document 2: Evidence Gathering Report**
- A report listing the details of all the anomalous conditions that are awaiting grading (ones that have been recommended for grading but not graded yet)
- Serves as a hand-off document from the generalization task for the evidence gathering activity, data mining personnel use this report to structure their research

**Document 3: System Type List**
- A list of the anomalous conditions that are awaiting grading broken up by their respective system type (APA practitioners can select which system to display using DAnGERS interface)
- This report is particularly useful for scheduling attendance of system experts during grading sessions, and communicates to management the volume of anomalous conditions to be graded

**Document 4: Low Detail Report**
- A summary report communicating the key APA details of all graded anomalous conditions
- Useful for providing a concise overview of current APA grading activities

**Document 5: High Detail Report**
- Builds on document 4, capturing the more detailed analysis of the grading activity
- The High Detail report contains all the details for a selected Generalization Group. The material contained in this report is specific to the Generalization Group selected using the DAnGERS user-interface
- Includes details of all observed anomalies and anomalous conditions within the generalization group, including evidence used for assigning FCI and CCI

# 5 Application to Other Mission Classes

Throughout the development phases of precursor analysis the method has been applied to Earth-to-Orbit transportation systems and crewed orbital science platforms; however the process in principle is valid for other mission classes (e.g., crewed and uncrewed orbital platforms, crewed lunar and planetary outposts, deep-space robotic missions, and other human space exploration missions). One category of mission class that is currently in operation at NASA that may benefit from a tailored derivative of precursor analysis is robotic space exploration.

The principle objective of precursor analysis is applicable for both crewed and uncrewed systems; assessing the history of operational anomalies to identify which anomaly failure mechanisms could potentially manifest themselves as problems in the future. Much can be learned from anomalies occurring on robotic platforms just as those onboard crewed vehicles; they have the potential to portend the future. Anomalies onboard uncrewed systems are significant to the extent that they either imply unexpected (out of nominal range) performance levels of well-understood elements or reveal risks that were misunderstood prior to the anomaly occurrence. For non-human related missions, the objective of a precursor analysis would focus on identifying underappreciated risks to system reliability rather than safety.

Uncrewed robotic exploration missions represent fundamentally different challenges from those crewed systems that precursor analysis has currently been applied to. Robotic missions entail a diverse range of functional systems (e.g. remote sensing, geological sampling, atmospheric entry descent & landing, surface roving); they operate in a wide variety of environmental conditions (e.g. deep space, atmospheric pressures and weather, radiation belts); the mission timeframes can be comparatively long and the inability to physically interact with the system (being able to repair any failure mechanisms) precludes many in-situ corrective actions when anomalies are encountered during a mission. Typically, when an anomaly is discovered during a robotic mission the corrective actions available are to: use as-is, update software on either ground or flight systems, or procedural modifications. Since the safety and reliability cycle of identify, fix, and fly (as traditionally applied to the STS and ISS programs) is not available for robotic missions, precursor analysis would have to take a different tack to identifying and addressing potential precursors. Rather than limiting the analysis to a single system or mission, operational anomalies would have to be used to identify failure mechanisms that could portend the potential for more severe consequences within an altogether different system or mission. Although robotic missions involve an array of different systems and subsystems, they are often designed to perform similar functions or operate in similar environments. For example, an anomaly that occurs within a Mars rover due to an atmospheric failure mechanism (dust ingress) is applicable to any other robotic system designed to operate within the same environment. Even though the scope for generalizing failure mechanisms within each individual robotic system may be significantly limited when compared to the STS or ISS, the APA process of generalizing failure mechanisms to other circumstances may be beneficial to robotic missions in an institutional sense.

This technique of analyzing anomalies across a range of historical missions may already be carried out in current risk management practices; however precursor analysis may provide an alternative method of structuring anomaly analysis across missions.

As discussed, the main requirement from a program in order to conduct precursor analysis is a source of operational anomalies with a means of identifying their causal mechanisms. NASA robotic missions currently report and record anomalous events and flight task data in electronic databases. If the practice is comparable to anomaly reporting and recording processes employed within the STS and ISS program (both of which have yielded suitable data sources to initiate precursor analysis) and depending on the level of information recorded in such anomaly reports, then data sources such as this could provide suitable caseloads for robotic cross-mission precursor analysis.

The typical practice of continuously analyzing anomalies through APA on a cyclical basis may not be the most appropriate review method for robotic missions, as so many anomalies are very specific to the hardware, conditions and the environment in which they occur and their causal mechanisms may not be applicable to other missions. Instead a triggering mechanism for reviewing records might be more appropriate, where anomalies occurring in systems with equivalent hardware/operating conditions onboard other missions (both current missions and future ones in the design phase) are flagged for precursor analysis. Details of such a mechanism could be developed in further studies into precursor analysis for other mission types.

# 6 Conclusion

The precursor analysis process described in this document is designed to be a practical, implementable means of identifying underappreciated sources of risk in NASA space systems, as evidenced by anomalous events that occur during the operation of the system. Once these sources of risk are identified, they can be brought into the overall system risk model, maintaining a current and consistent knowledge of risk that uses all available information. In some cases, design or operating practice will be adjusted to mitigate to the newly realized risk. Either way, a pathway is provided to reduce the gap between model and reality in an iterative and continuous fashion. Maintaining this correspondence between model and reality is conducive to sound risk-informed decision-making. Without it, decisions may be made that unknowingly accept a significant risk based on little or no actual evidence of the prevalence for the underlying conditions.

The process implements a graded approach to analysis, using screening and qualitative grading in order to focus subsequent monitoring and analytical attention only on those potential failure situations that warrant it, based on their risk grading (PPI). Anomalous conditions with low PPIs are recommended for no further action; those with moderate PPIs are recommended for continued observation and trending; and those with high PPIs are recommended for explicit risk modeling. Risk modeling includes scenario-based modeling techniques to identify specific potential engineering vulnerabilities that can be addressed by subsequent testing, analysis, design change, or procedural modification. As such, it leverages existing anomaly data in a manner that provides a maximum potential for systemic risk reduction.

Although grading employs qualitative analysis, the rationales used to assign the PPI are expected to be grounded in solid evidence. When the evidence is weak, the grading is biased conservatively (toward more observation or analysis rather than less). The conservatism of the grading increases as the evidence caliber decreases, assuring that situations with large uncertainty are handled conservatively. This ties the process to high-confidence estimates as opposed to mean values.

To date, the APA process has been applied to two NASA programs, STS (both Orbiter and SRB) and the ISS. Over time, the process has evolved to adapt and integrate with the specific requirements of each program's risk management practices. To illustrate one of these adaptations, in pilot applications to the STS program APA was only concerned with 1 severe consequence, LOC/LOV (at the direction of program risk management) and the generalization process only took into account anomalous conditions that could result in LOC/LOV. APA was later applied to the ISS program, risk management practices within this program focus on Loss of Science (LOS) and Loss of Mission (LOM) consequences in addition to LOC/LOV. In the interest of capturing anomalous conditions that could potentially result in LOS and LOM, the process was refined to allow consideration of multiple severe consequences during generalization. Each step in the process can be tailored in various ways so that APA provides the best possible method of identifying potential precursors within the system.

Up to this point methods and processes for evaluating and grading anomalous conditions, as well as modeling and quantifying their risk significance, have been presented, but in the end the process, by its very name, implies the identification of accident precursors. However, the fact of the matter is that the process, in and of its self, does not designate precursors. This is because an accident precursor is defined not so much by the absolute values of the risk significance measures but instead by the way that those risk significance measures are interpreted by a program. Thus, it is the responsibility of the program, if it so chooses, to define criteria by which to designate an anomalous condition (whether actually observed or postulated) as an accident precursor.

For example, a program may choose to define as precursors those anomalies in any generalization group containing an anomalous condition with a risk measurement (ACRI) at or above a program-defined threshold.

It is important that once the criteria are defined, they be adhered to, since not doing so relegates the program to the subjective type of attention given to anomalies that APA is attempting to remedy. Once an anomalous condition has been designated as an accident precursor, risk modeling (PVI) provides a means of focusing program resources upon the parameters and system states that either require more information or are explicitly driving the failure propagation to severe consequences.

# Appendix A - Acronyms

ACRI – Anomaly Condition Risk Importance
APA – Accident Precursor Analysis
ASAP – Aerospace Safety Advisory Panel
ASP – Accident Sequence Precursor
BCDU – Battery Charge/Discharge Unit
CCI – Conditional Consequence Index
CDR – Critical Design Review
DAnGERS – Deliberative Anomaly Grading & Evaluation for Risk Significance
DC – Data Caliber
EC – Evidence Caliber
EMU – Extravehicular Mobility Unit
ET – External Tank
EVA – Extravehicular Activity
FCI – Failure Condition Index
FCOC – Failure Condition of Concern
FESD – Functional Event Sequence Diagram
FMEA – Failure Modes and Effects Analysis
HAZOP – Hazard and Operability
IFA – In-Flight Anomalies
IFI – Item For Investigation
ISS – International Space Station
JSC – Johnson Space Center
LDI – Local Data Interface
LOC – Loss of Crew
LOM – Loss of Mission
LOS – Loss of Science
LOV – Loss of Vehicle
MAS – Mission Assurance System
MER – Mission Evaluation Room
NLGD – Nose Landing Gear Door
NPR – NASA Procedural Requirements
NRC – Nuclear Regulatory Commission
OMS – Orbital Maneuvering System
OSMA – Office of Safety and Mission Assurance
PDR – Preliminary Design Review
PPI – Potential Problem Index
PRA – Probabilistic Risk Analysis
PRACA – Problem Reporting and Corrective Action
PRCB – Program Requirements Control Board
RCC – Reinforced Carbon-Carbon
RCS – Reaction Control System
S&MA – Safety and Mission Assurance
SRB – Solid Rocket Booster

SSU – Sequential Shunt Unit
STS – Space Transportation System
TPS – Thermal Protection System

## Appendix B - References

1. "Three Mile Island: A Report to the Commissioners and to the Public," NUREG/CR-1250, Mitchell Rogovin, Director, Nuclear Regulatory Commission Special Inquiry Group (NRC, 1980). [See esp. p. 94 of Vol. 1]

2. Presidential Commission on the Space Shuttle Challenger Accident, "Report of the Presidential Commission on the Space Shuttle Challenger Accident" (Rogers Commission Report), 1986.

3. Columbia Accident Investigation Board, "Columbia Accident Investigation Board Report," 2003.

4. NASA Aerospace Safety Advisory Panel, "Annual Report for 2002," 2003.

5. NASA Aerospace Safety Advisory Panel, "Annual Report for 2006," 2007.

6. NASA Office of Safety and Mission Assurance, "NASA System Safety Handbook", NASA/SP-2010-580, Washington, DC, 2011.

7. Sattison M.B., "Nuclear Accident Precursor Assessment: The Accident Sequence Precursor Program," appearing in J.R. Phimister, V.M. Bier, and H.C. Kunreuther, eds., "Accident Precursor Analysis and Management: Reducing Technological Risk Through Diligence," National Academy of Engineering of the National Academies (The National Academies Press, Washington, DC, 2003).

8. Vesely W., Stamatelatos M., "Risk-Based Anomaly Precursor Analysis," presented at Space Systems Engineering and Risk Management Symposium, 2008.

9. Vesely W., "Anomaly Precursor Analysis: Concepts, Approaches, and Examples," 2007

10. Maggio G., Youngblood R., Everett C., and Hall A., "An Accident Precursor Analysis Technical Approach for NASA Space Systems and Pilot Program Application," NASA OSMA Technical Report, 2008.

11. Groen F., Lutomski M., "NASA's Accident Precursor Analysis Process and the International Space Station," presented at the Trilateral Safety and Mission Assurance Conference, NASA Safety Center, Cleveland, OH. 2010.

12. Reason J., "Human Error," Cambridge University Press, 1990.

13. NASA Office of Safety and Mission Assurance, "The Davis-Besse Close Call,", System Failure Case Study, 2006.

14. NASA Johnson Space Center, "Problem Reporting and Corrective Action (PRACA) System Requirements," NSTS 08126, 2000.

15. NASA Office of Safety and Mission Assurance, "NASA Root Cause Analysis Supplemental Training Material," Washington, DC., 2008.

16. NASA Johnson Space Center, "International Space Station Program Risk Management Plan," SSP 50175 Rev B-1, 2007.

17. U.S. Nuclear Regulatory Commission, "Measures of Risk Importance and their Applications", NUREG/CR-3385, Washington, DC, 1983.

18. NASA Office of Safety and Mission Assurance, "Probabilistic Risk Assessment Procedures Guide for NASA Managers and Practitioners", NASA/SP-2011-3421, Washington, DC, 2011.

19. National Institute of Standards and Technology, "NIST/SEMATECH e-Handbook of Statistical Methods," http://www.itl.nist.gov/div898/handbook/

20. NASA. "Bayesian Inference for NASA Probabilistic Risk and Reliability Analysis," NASA/SP 2009 569. Washington, DC. 2009.

# Appendix C - The Tool for Deliberative Anomaly Grading and Evaluation of Risk Significance (DAnGERS)

DAnGERS is a software tool that is available to NASA to support generalization and grading exercises. The tool shown herein was developed for the ISS program and field-tested during numerous pilot applications (a Space Shuttle variant of DAnGERS also exists). It serves as a representative example of how to apply the process in a structured application.

The tool supports generalization of observed anomalies and grading of anomalous conditions by structuring and documenting the discussion of each anomaly record, including the evidentiary basis for the findings of the discussion. When sufficient data has been collected and input for a given anomalous condition, DAnGERS uses a series of metrics and calculations to automatically assign a "Potential Problem Index" (PPI) for it, which is used to generate a recommendation for further analysis.

The following sections provide a step-by step overview of how to operate and complete DAnGERS during precursor analysis exercises. The structure of this guide corresponds to the flow of the tool, and is designed to be operated in two working-session environments; **Anomaly Review & Generalization**, and **Grading**.

Examples shown in this appendix are notional, though derived from events that occurred on the ISS.

## C.1    DAnGERS Main Menu

The main menu for the tool enables the user to launch various modules of the tool. Five options are displayed; 'Import Data', 'Generalization', 'Grading', 'Reports' and 'Upload to MAS'. Clicking on any of these buttons will launch a separate form that takes the user through the steps to complete each process.

DAnGERS requires a caseload of anomalous events to be imported in order to evaluate that data using the APA process. The 'Import Data' module allows records to be added to the database within the tool, a caseload of anomaly records can either be created manually (using MS Excel) or existing anomaly reporting tools can be utilized to generate caseloads and imported into DAnGERS. Once the tool contains a caseload ready for evaluation, the button labeled 'Generalization' will launch the user interface for the generalization process step, displaying the Anomaly Review page. The 'Grading' button launches the module used for the grading process step, to be used following evidence gathering. The 'Reports' button launches a menu which shows all of the available methods of report generation, these may be generated at any time throughout an APA cycle, and a variety of reports are available to reflect the appropriate level of detail desired. Finally, the 'Upload to MAS' button launches a module that uploads the information contained within DAnGERS to a prototype precursor information repository within the ISS Mission Assurance System (MAS), an online NASA data management tool that enables NASA personnel to remotely view, track, and download Safety and Mission Assurance related information.

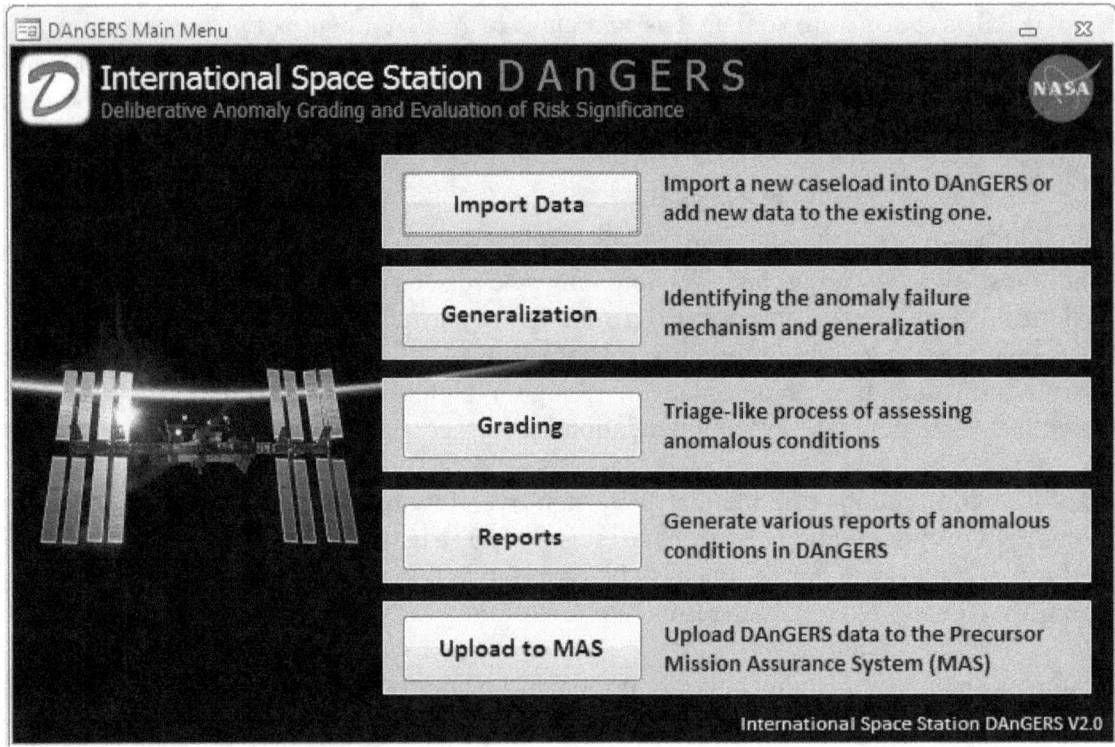

**Figure C-1 ISS DAnGERS Main Menu**

## C.2     Generalization

Once a caseload has been imported into the tool the 'Generalization' form can be launched, and will display the first tab of the generalization form, the 'Anomaly Review' tab. This page shows caseload data for each anomaly as shown in the Anomaly Title field at the top of the page.

The data displayed within the grey area directly corresponds to the caseload data imported into the tool (Figure C-2); it provides sufficient information to characterize the anomaly scenario for the group, so that each member has a clear understanding of the anomaly for which they are reviewing.

Once the anomaly information has been reviewed by the group, they will discuss the identification of the anomaly failure mechanism. A list of example failure mechanisms is available via a drop-down menu; these have been generated through the experience of numerous APA working sessions and provide a useful 'go-to' list for the generalization team to consider when deliberating an appropriate failure mechanism. In some cases, the record of an anomaly may be insufficient to support the grading activity, as would be the case if a failure mechanism for the observed anomaly could not be identified. In such a case, the user may simply note the data insufficiency, record any remarks on the level of available information, and proceed to the next anomaly record. Consideration of records like these may be deferred pending additional information, and continued when that information becomes available.

Below the failure mechanism section there is a control for recording the operational phase of the program that the anomaly occurred under, the options are; On-Orbit, Ground Ops, or Other. This field is present for differentiating certain activities (testing) that may result in anomalies occurring outside of the ISS operational envelope.

The final data collected on the Anomaly Review page records the System, Subsystem, Flight Element and component in which the observed anomaly occurred (Figure C-2). A detailed taxonomy of the ISS architecture is contained within the tool, the DAnGERS operator can select the appropriate location of the anomaly from the drop-down menus. This procedure of identifying the location of the observed anomaly is helpful for characterizing the anomaly, as well as setting the stage for the generalization process that will follow.

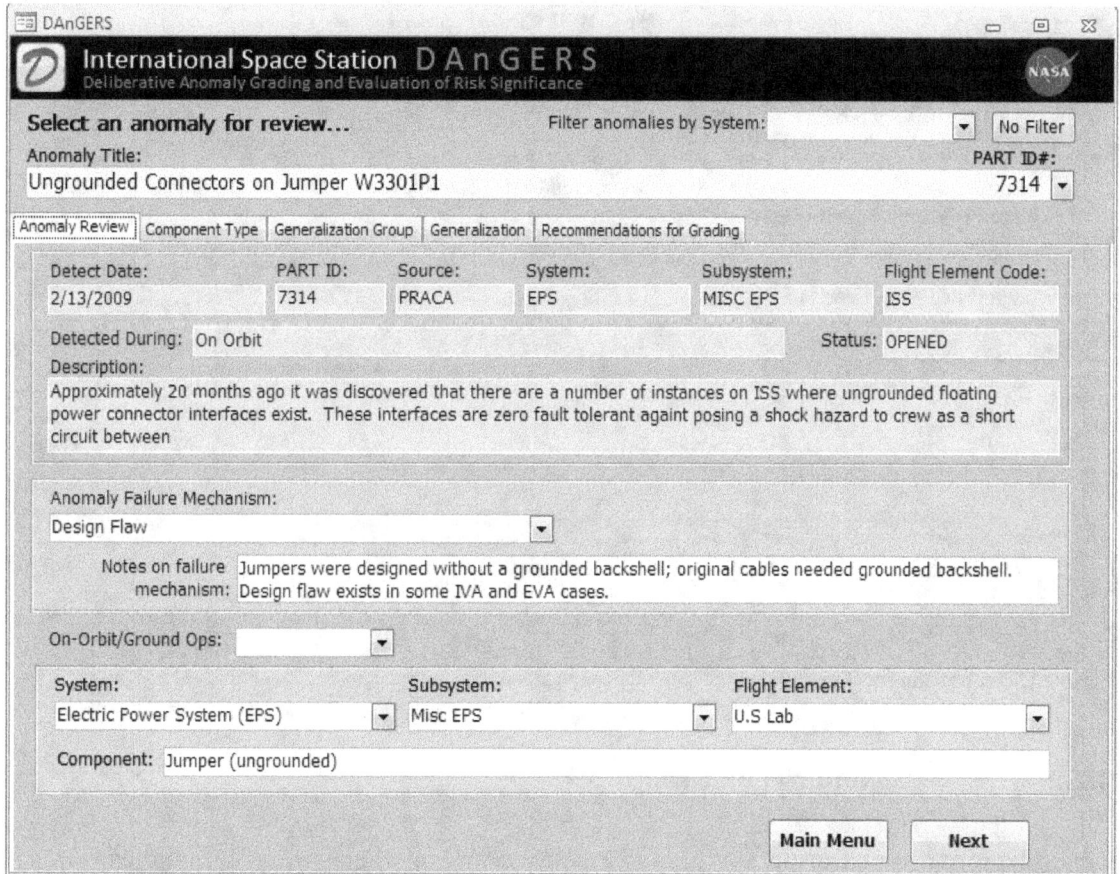

**Figure C-2 Anomaly Review Page**

### C.2.1 Component Type

The next step of the generalization process identifies the generic component type. Generalization is constrained to the generic component type that experienced the observed anomaly. The Generalization team will discuss and determine the generic component type within the context of the observed anomaly.

The Generic Component Type field is a dynamic drop-down menu, this means that the user can click on the arrow and see a list of generic component types or they can type in their own generic component type into the text box. New entries that are typed-in are added to the list so that it will be visible in the drop-down menu in future. This means that the first time the tool is used the list will be blank, but the list will grow as new generic component types are identified.

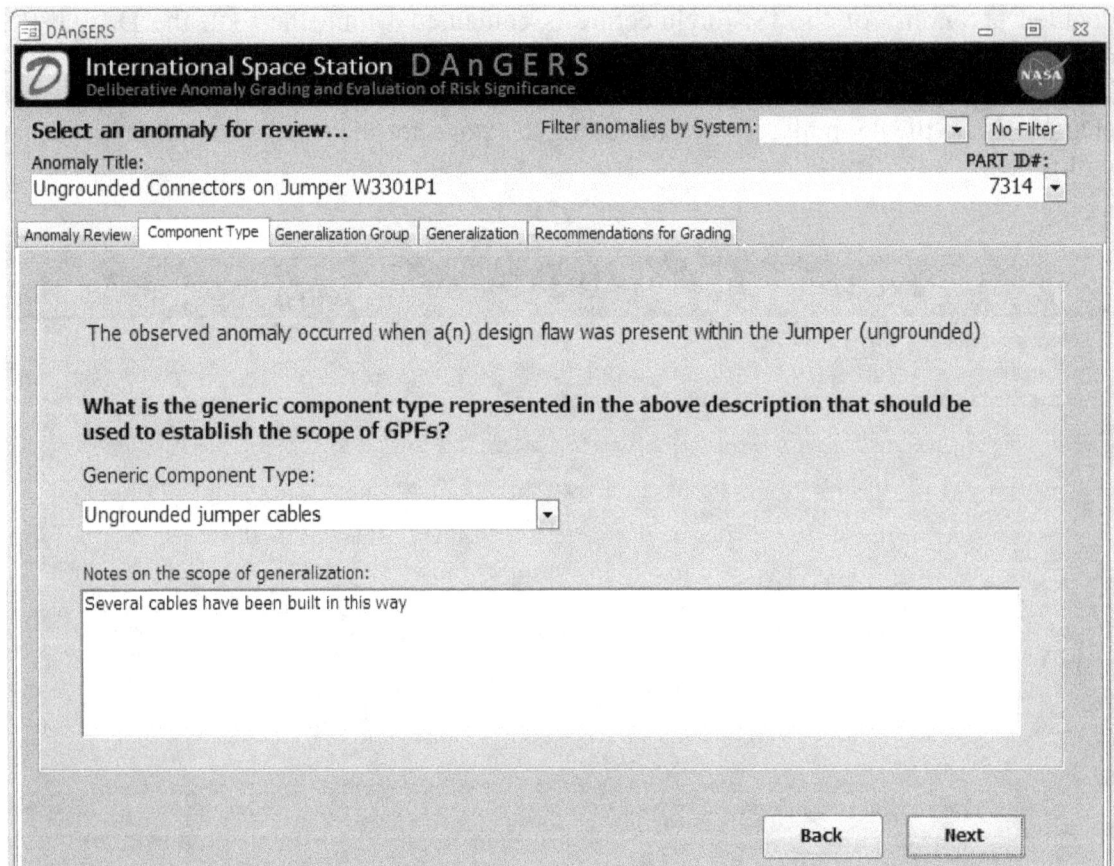

Figure C-3 Component Type Page

### C.2.2 Generalization Group

Generalization groups are a useful way of grouping together similar anomalies. On the Generalization Group page, the tool will display any groups that have a matching failure mechanism and component type to the observed anomaly. If there are no groups displayed on this page, it is because there have been no previous anomalies that match this criteria (Shown in Figure C-4) In this case, a new generalization group is be created and assigned to the observed anomaly by the operator.

If the observed anomaly's failure mechanism and component type match an existing generalization group within the tool, then the matching group will be displayed as soon as the page is visible. In this case the team will review the information displayed regarding the failure mechanism and component type, if the observed anomaly is deemed to 'fit' within the generalization group, the DAnGERS operator will assign it to the group. If the generalization group displayed does not adequately match the observed anomaly, then the 'Create New' button can be used to create a new group.

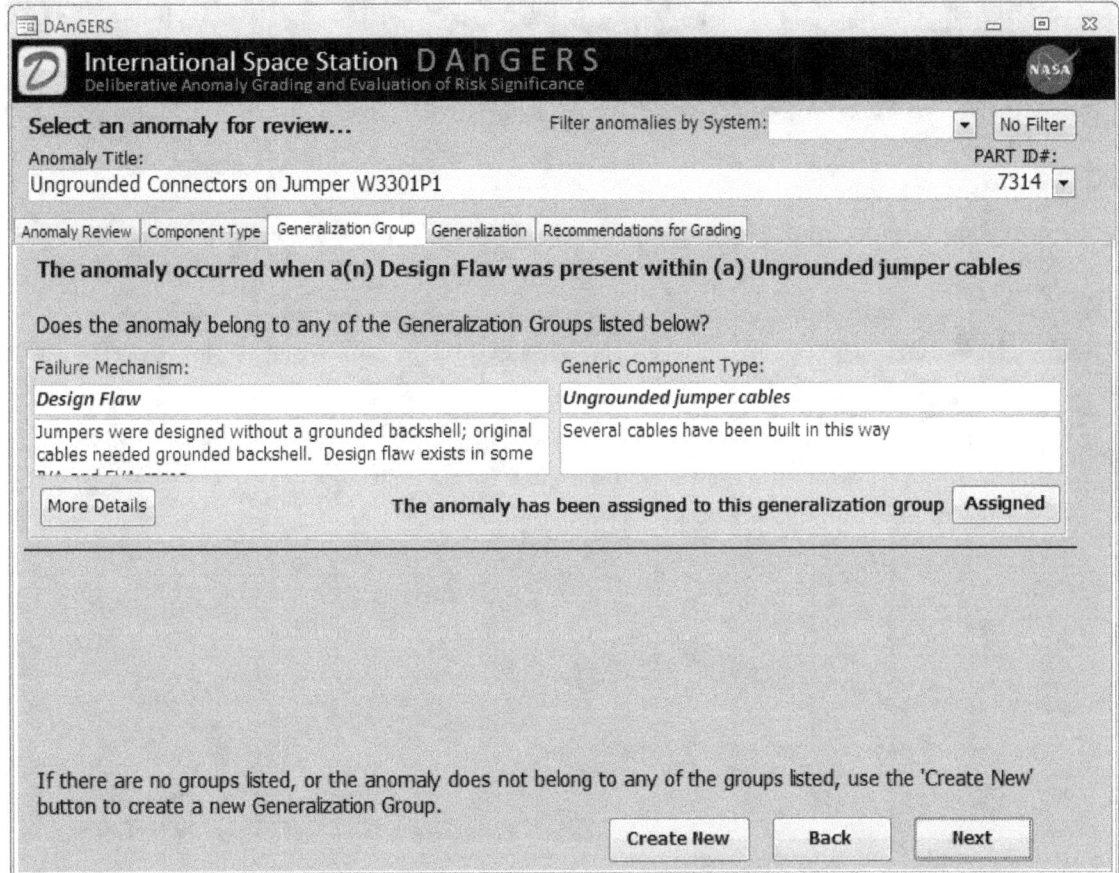

Figure C-4 Generalization Group - New Group

### C.2.3 Generalization

Each component within the scope of the generic component type is evaluated to identify other locations or time frames that are susceptible to the same failure mechanism, and which could potentially lead to more severe consequences.

The question at the top of page ("In what other…." Figure C-5) directs the team members to generalize the failure mechanism to other components within the generic component type. Additionally, the team is asked to consider other activities under which the failure mechanism could lead to more severe consequences.

Information displayed in the blue form corresponds to the observed anomaly data entered on the 'Anomaly Review' page. At this step, the component in the observed anomaly is known, however the team must identify the timeframe in which the observed anomaly occurred (and could lead to more severe consequences), the failure condition of concern (FCOC) that could be created, and the severe consequence that potentially would have resulted from the accident progression.

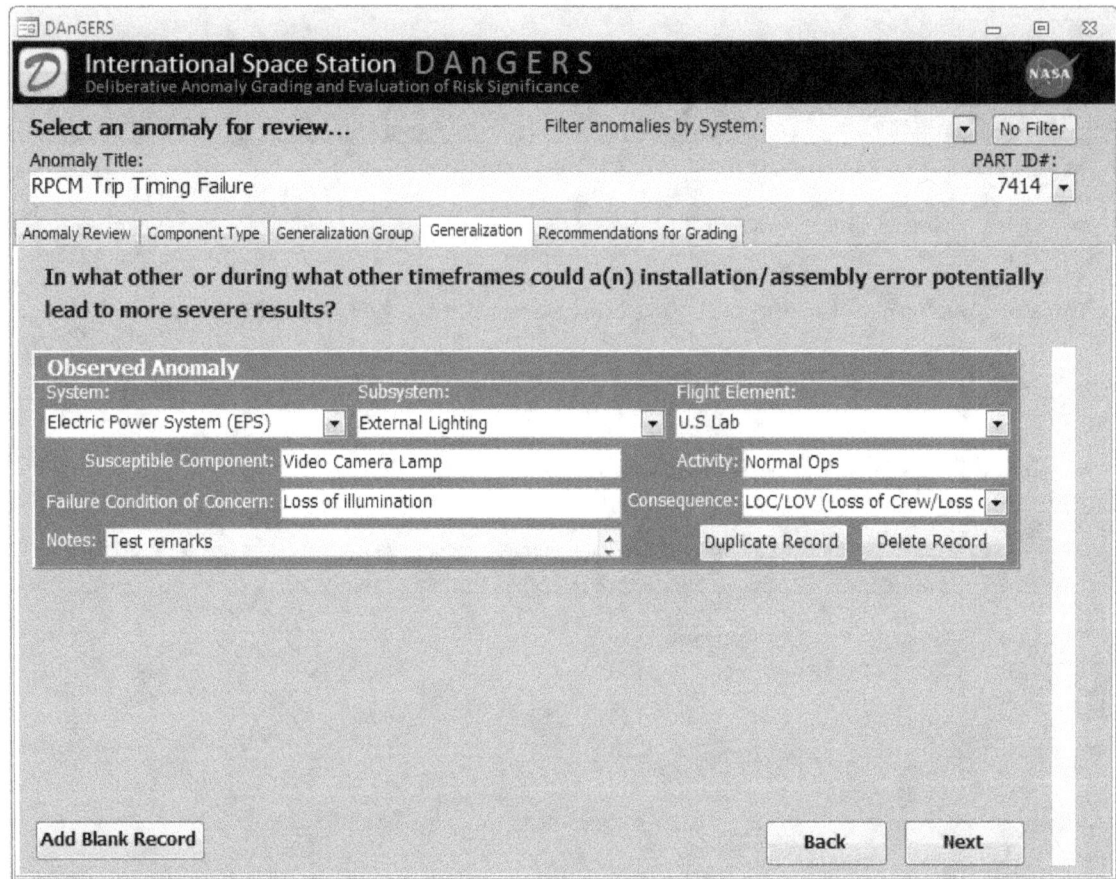

**Figure C-5 Sample Generalization Page 1**

As the group begins to generalize the anomaly failure mechanism to other locations and times, the DAnGERS operator will correspondingly create new anomalous conditions with varying susceptible components and activities until all potentially more severe anomalous conditions are recorded. A generalization team may wish to assess a scenario for its potential to produce two severe consequences; this can be accomplished by duplicating the record and selecting the two different options for severe consequence.

The goal at this point is to determine the scope of anomalous conditions that could be forwarded to the grading exercise, making efforts to identify all locations and time frames where the anomaly failure mechanism has the potential to lead to severe consequences, while also minimizing the number of ultimately insignificant anomalous conditions.

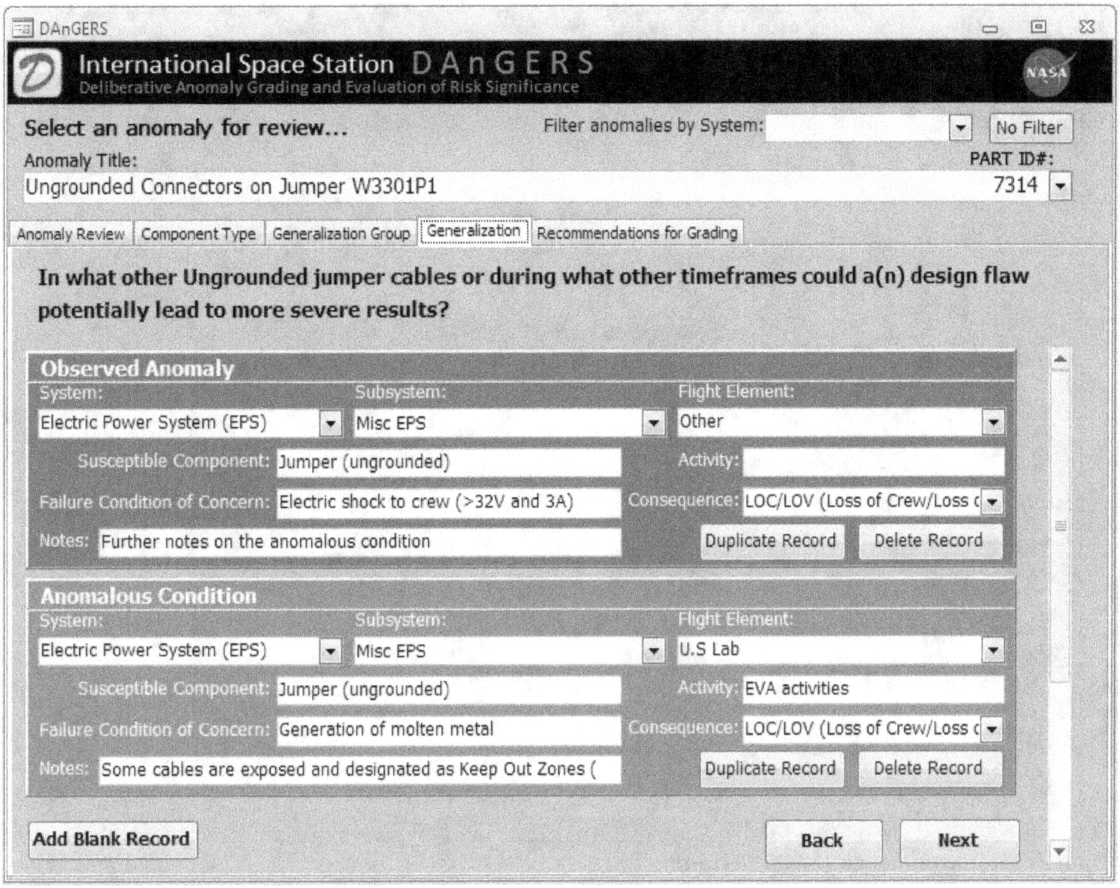

Figure C-6 Completed Generalization Page

### C.2.4 Recommendations for Grading

This page of the tool displays a summary of all the current data generated from the generalization process, to provide a picture of the scenario that each anomalous condition presents. The team will deliberate on which anomalous conditions should be pursued for Grading, once a conclusion is reached the operator will use the Yes/No radio buttons located to the right of each anomalous condition to recommend a subset of anomalous conditions for grading. Those that are toggled to 'No' will NOT be carried on any further in the precursor analysis process.

The output of generalization is a set of anomalous conditions deemed worthy of further evaluation and grading. Prior to engaging in grading, data on each of the affected subsystems needs to be collected, as well as sufficient system-level data to evaluate the potential for each failure condition of concern to occur, given the presence of the failure mechanism, and the potential for the failure condition of concern to propagate to severe consequences.

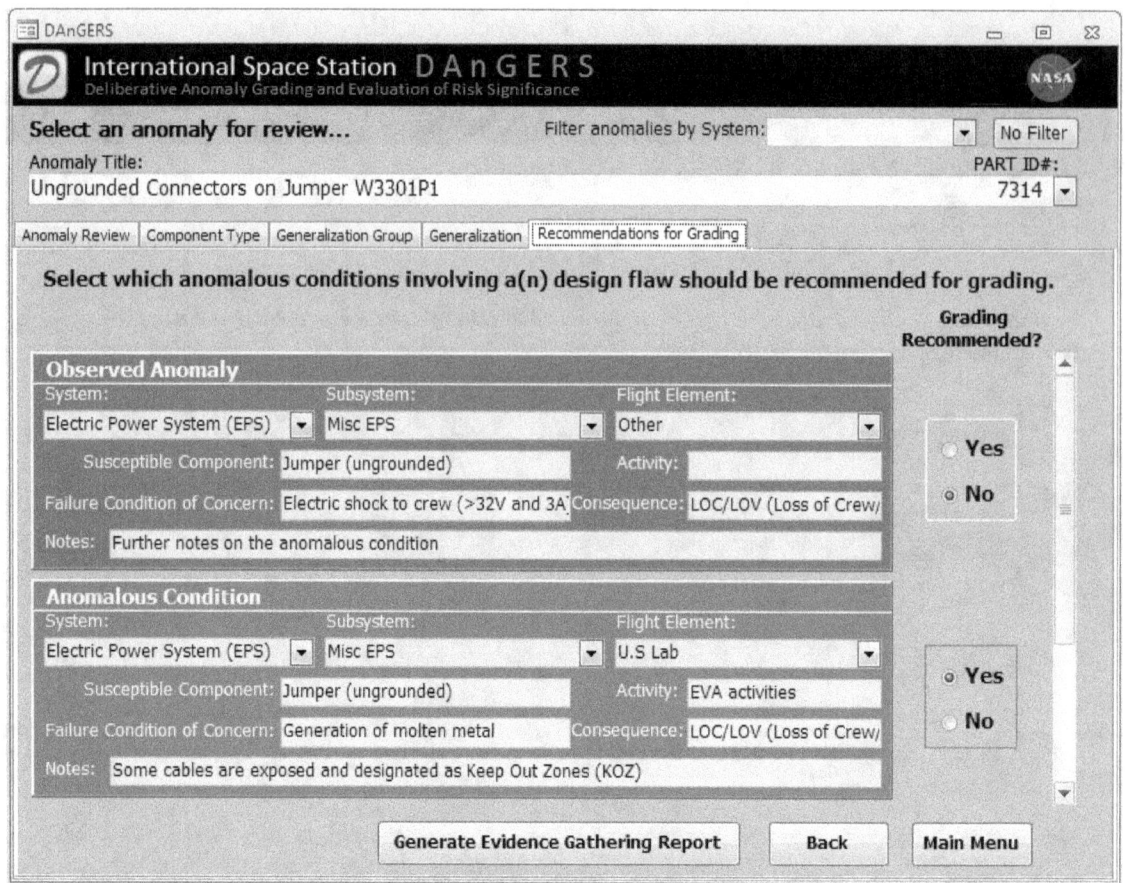

**Figure C-7 Recommendations for Grading Page**

The last action in the Generalization process is to produce a report of all the pertinent data collected so far that will be provided for the evidence gathering personnel. This report is produced by DAnGERS automatically by clicking on the 'Generate Evidence Gathering Report' button (See Appendix D for further details on Reports).

## C.3 Grading

When sufficient evidence has been gathered for those anomalous conditions which have been recommended for grading, a grading session is scheduled with a team of experts to review and assess the potential for problems created by each anomalous condition. To begin the grading step, the user must select the 'Grading" button from the main menu to launch the grading module.

### C.3.1 Initial Review

The first page of the grading module presents an overview of the current anomalous condition to be graded. It may be days, weeks, or longer since the generalization session took place which characterized a given anomalous condition, and this page serves to refresh the group about the anomalous condition (Figure C-8).

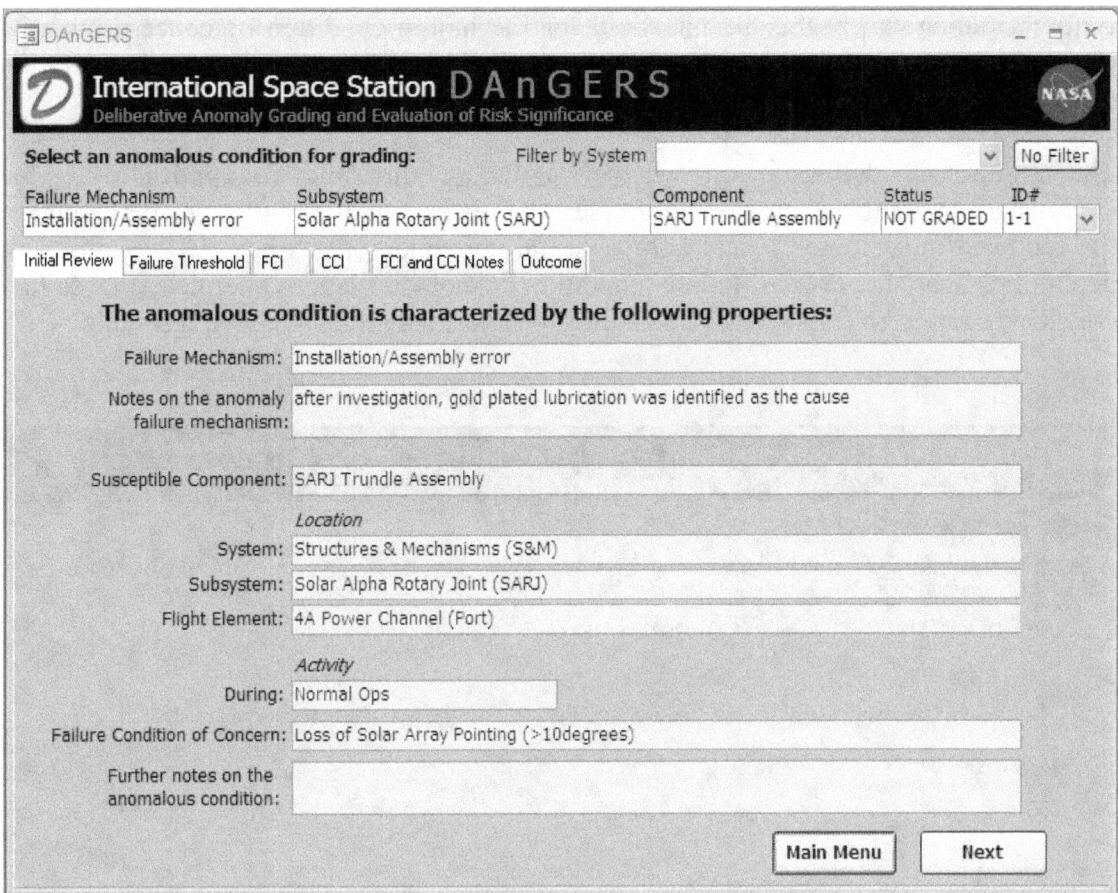

Figure C-8 Initial Review of Anomalous Condition

Navigation of the grading module is similar to the generalization module; however the 'Anomaly Title' drop down is replaced with a drop down for selecting anomalous conditions. At this stage, the process is no longer concerned with observed anomalies, and instead only considers anomalous conditions. This drop-down menu uses the combination of failure mechanism, subsystem, and component to characterize each anomalous condition and facilitate navigation. Additionally, it includes a 'Status' indicator to designate if the anomalous condition has been graded yet or not. This can be useful during a grading session to identify records that have already been evaluated. Finally, a 'Filter by System' drop-down is provided as a means to limit the anomalous conditions to only certain systems at a time. If many anomalous conditions exist for many different systems, yet a particular grading team only wishes to review those from ONE system, this feature should be used to limit the scope to anomalous conditions within that system.

## C.3.2 Refining the Failure Threshold

Before Grading can begin, the threshold for the failure condition of concern must be evaluated. The failure condition of concern is identified during the generalization process; however it may not explicitly define any threshold for failure to occur.

The group will deliberate on whether a failure threshold is applicable to the circumstances. As the discussion progresses, the operator will record notes pertaining to the rationale into the notes field below the failure condition textbox. If the consensus determines that the current failure condition is not sufficiently defined, the failure condition of concern displayed can be edited to reflect the group's agreed threshold.

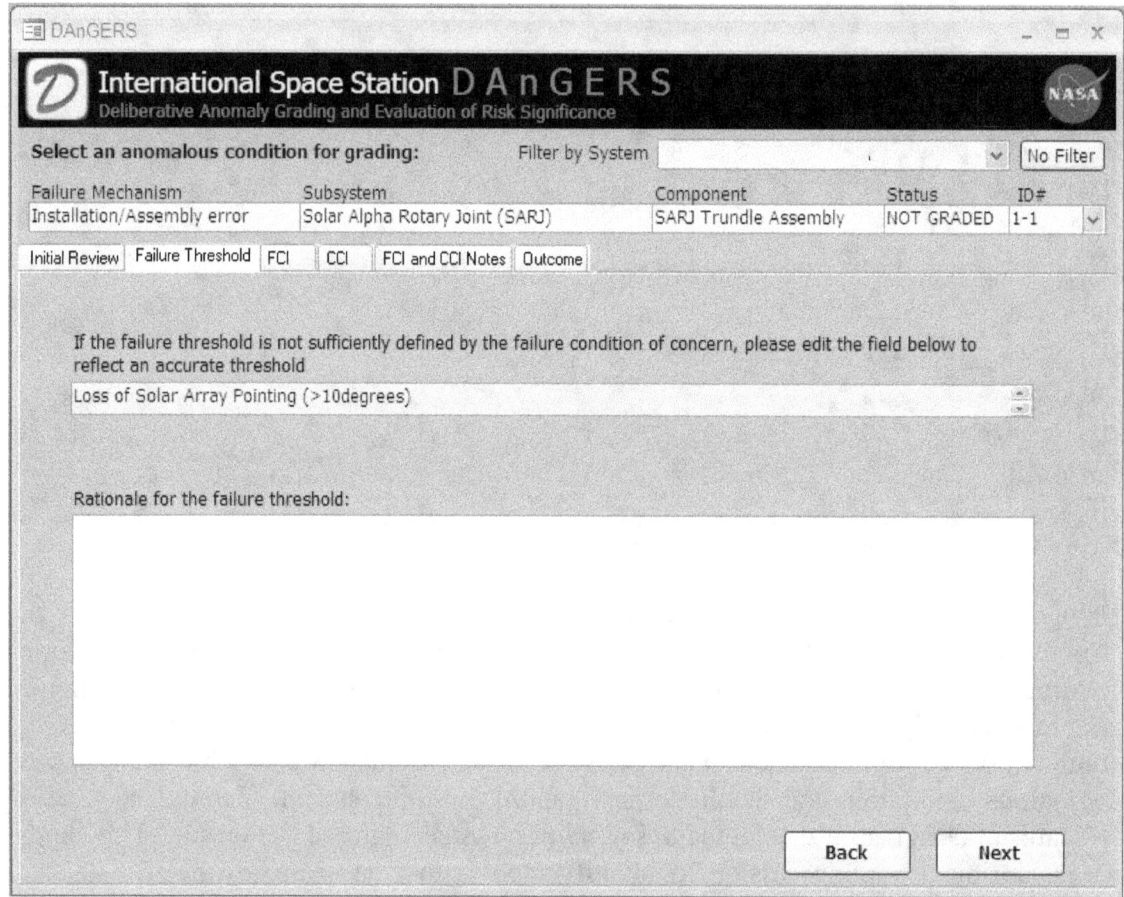

**Figure C-9 Failure Threshold**

## C.3.3 Failure Condition Index (FCI)

FCI is assigned based on the likelihood of the "failure condition of concern" occurring, given the occurrence of the failure mechanism in the susceptible component.

During evaluation of FCI, the grading team identifies the data upon which they are basing their FCI assignment, this data is provided by the evidence gathering exercise. The tool requires the user to record data relating to the evidence submitted, in order to incorporate the caliber of the supporting evidence. This begins with identifying the data type. Click on the 'Data Type' drop-down menu to see a list of relevant data types, and select one that matches the evidence (Figure C-10).

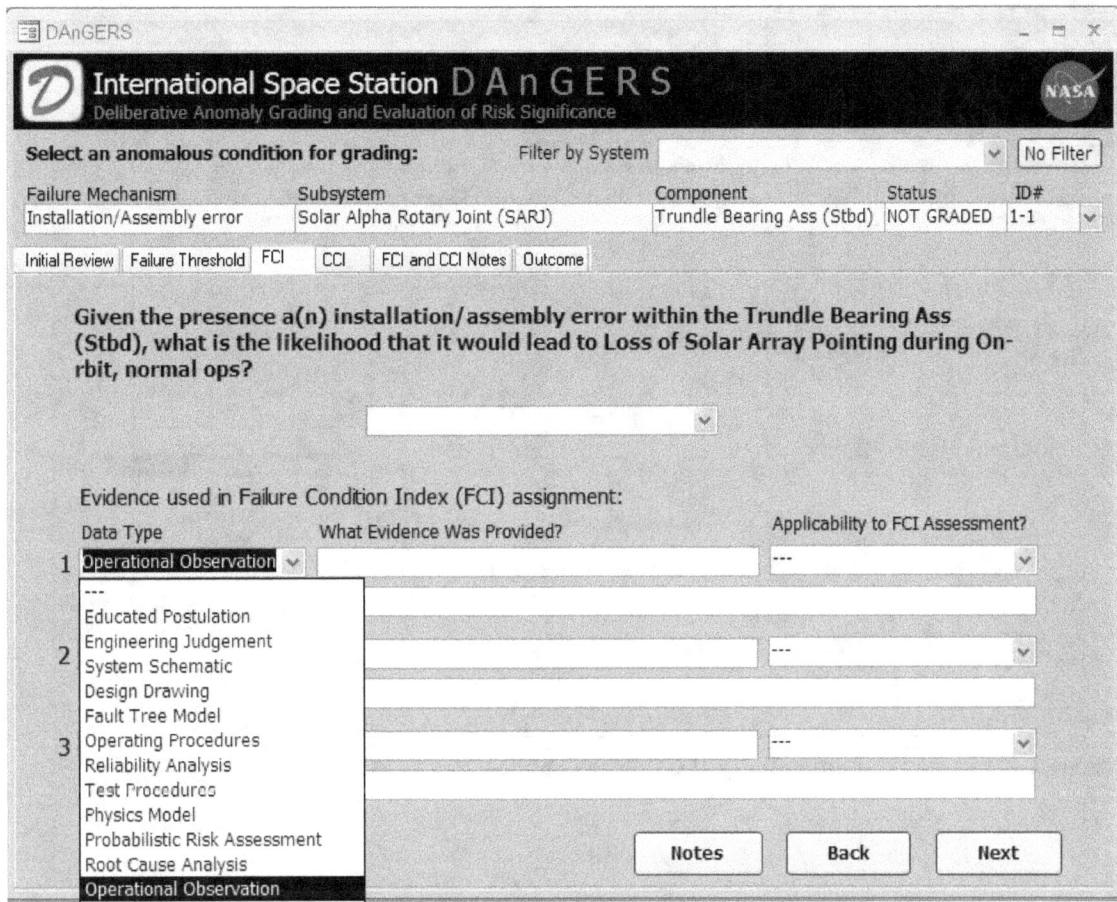

Figure C-10 Failure Condition Index (FCI) Page (Data Type)

Each piece of evidence must also be assigned an applicability indicating the degree to which the objective evidence addresses the potential for the anomalous condition to produce the failure condition of concern. The applicability is assigned as a qualitative descriptor using the drop-down menu labeled 'Applicability to FCI Assessment' (Figure C-11).

The final text box for FCI evidence recording requires the user to document the data source. This must be detailed enough so that a third party can easily locate this piece of data at a later date.

Up to three separate sources of evidence can be recorded in the tool, each time entering the appropriate data type, evidence, applicability, and data source. Further remarks regarding the FCI assignment, group rationale or any other pertinent information can be recorded in the notes section. (See Figure C-12).

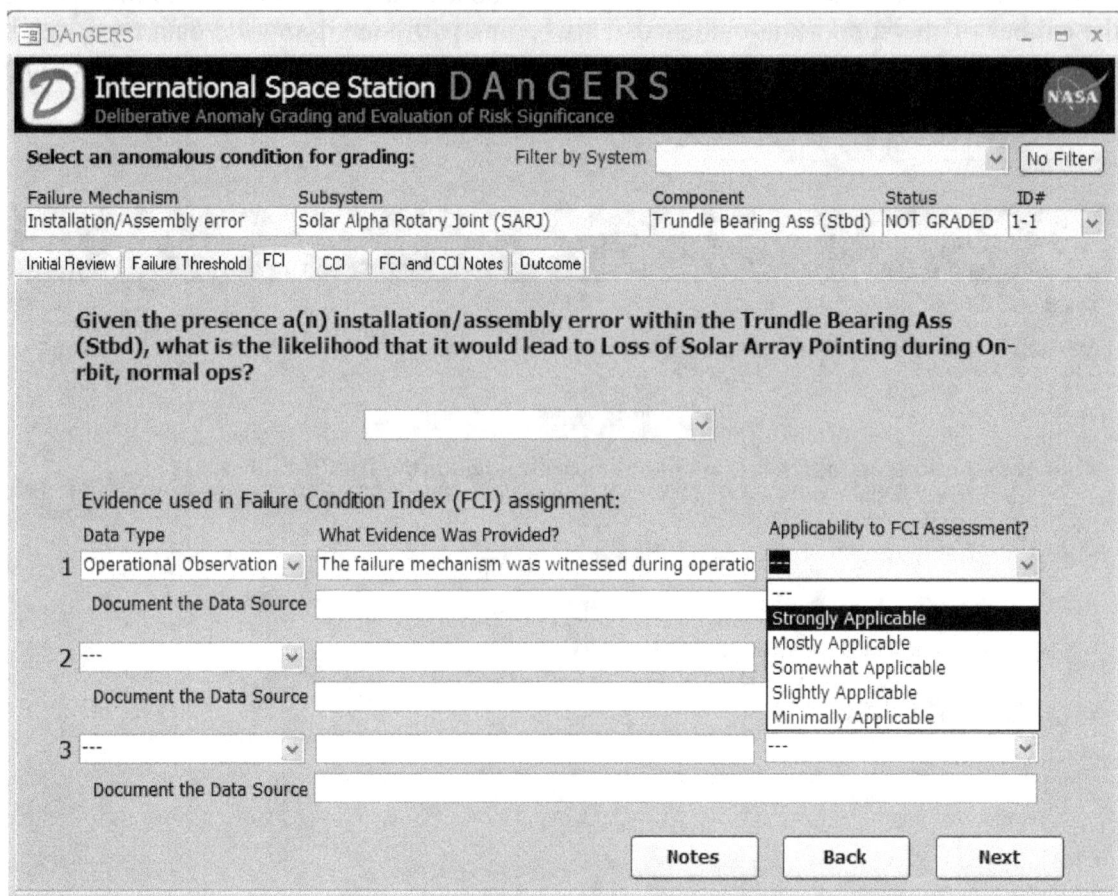

Figure C-11 Failure Condition Index (FCI) Page (Applicability to FCI)

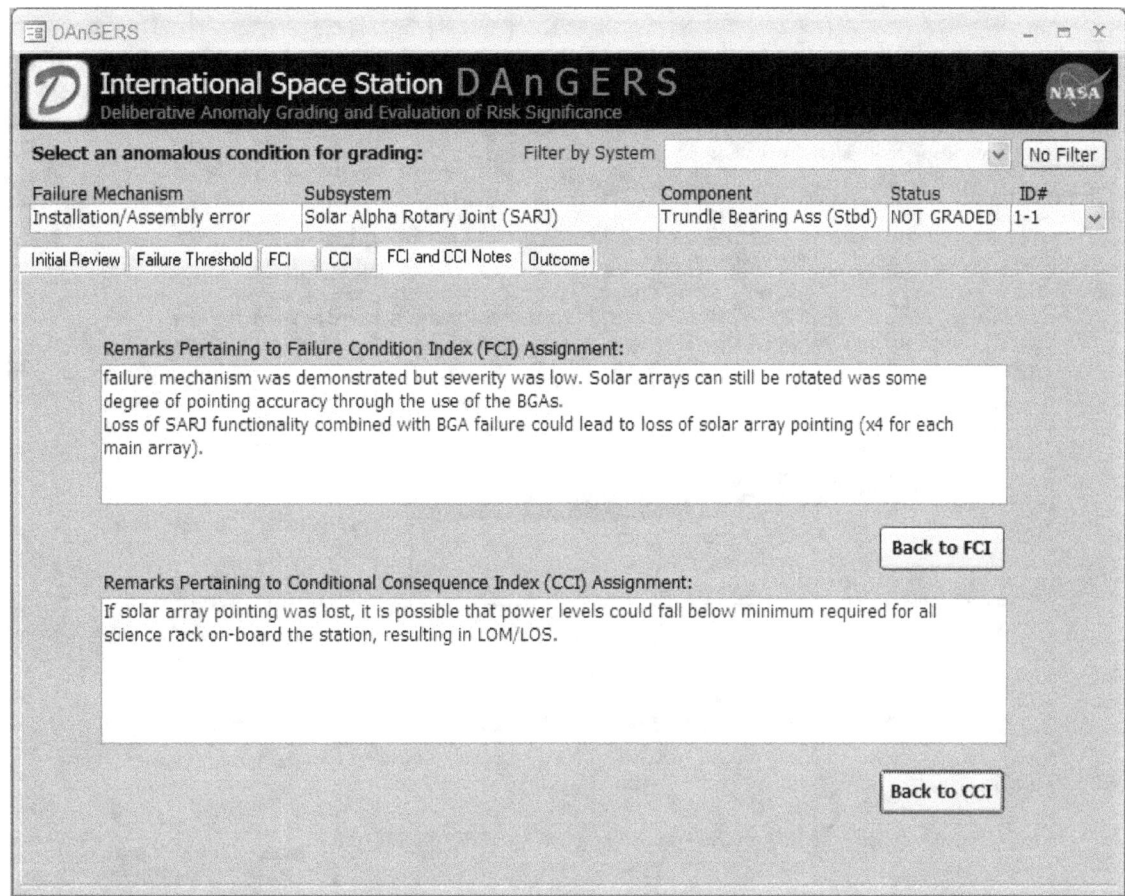

**Figure C-12 FCI and CCI Notes Page**

Once all the supporting evidence has been completed and documented the FCI can be assigned. Grading of the FCI is applied using a list of qualitative descriptors. A drop-down menu is used to select the most appropriate out of 6 possible qualitative FCI descriptors; Very Likely, Likely, Possible, Unlikely, Highly Unlikely, and Non-Credible (Figure C-13). The group will discuss the merits of each piece of evidence and the potential relevance to the assignment of FCI likelihood.

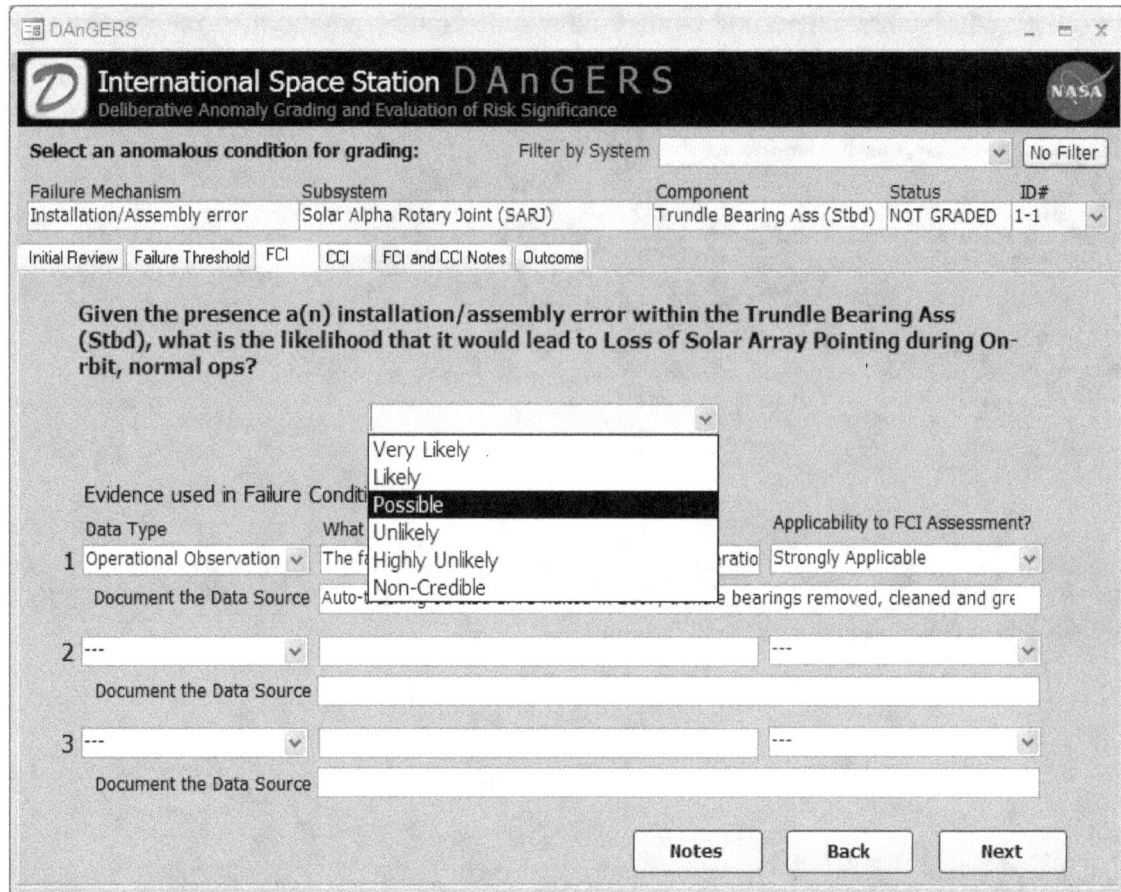

**Figure C-13 FCI Page (Qualitative Descriptors)**

The drop-down menu can be selected to display the list of descriptors during the group discussion, this will help guide the group to choose which best describes the likelihood of progression from failure mechanism to failure condition of concern.

### C.3.4 Conditional Consequence Index (CCI)

The Conditional Consequence Index (CCI) indicates the potential that the failure condition of concern will result in severe consequences. This is a function of the degree to which safeguards or barriers are present to mitigate severe consequences given the failure condition of concern. CCI is graded based on the same list of qualitative descriptors as the FCI.

The CCI page is representative of the same procedure for assigning an FCI, the design is very similar and they can be filled out much in the same way.

Like FCI, the assigned qualitative CCI descriptor must ultimately be based on evidence. Hence, the same process for assessing evidence applicability is used for CCI evaluation that was used for FCI evaluation. The fields can be completed in the same way as the FCI evidence; 'What Evidence Was Provided?', 'Applicability to CCI Assessment?', and 'Document the Data Source'.

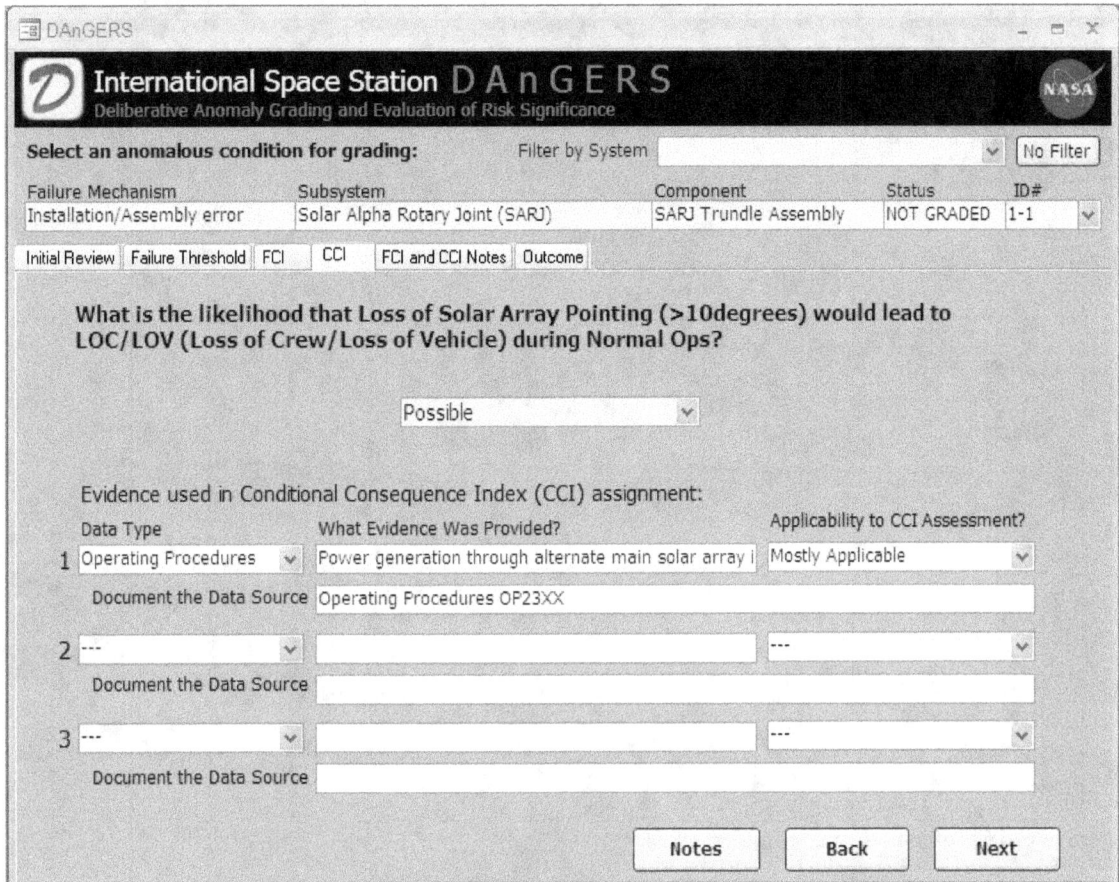

Figure C-14 Conditional Consequence Index (CCI) Page (Completed)

Grading of the CCI is applied using the same list of qualitative descriptors used for grading the FCI; Very Likely, Likely, Possible, Unlikely, Highly Unlikely, and Non-Credible. As practiced for FCI, follow the group discussion and select the agreed option. A completed sample CCI page is shown in Figure C-14.

### C.3.5　　　Outcome

The Outcome page displays the APA results from the grading process for the anomalous condition just evaluated. These results comprise of; the Potential Problem Index (PPI), the Evidence Caliber (EC), and Recommended Further Action (Figure C-15 shows a completed 'Outcome' page).

PPI is the metric used in the APA process to grade an anomalous condition for either risk modeling, observation and trending, or no further analysis. The Evidence Caliber (EC) is a function of the data type and assigned applicability of each piece of supporting data used by the grading team when assigning FCI and CCI.

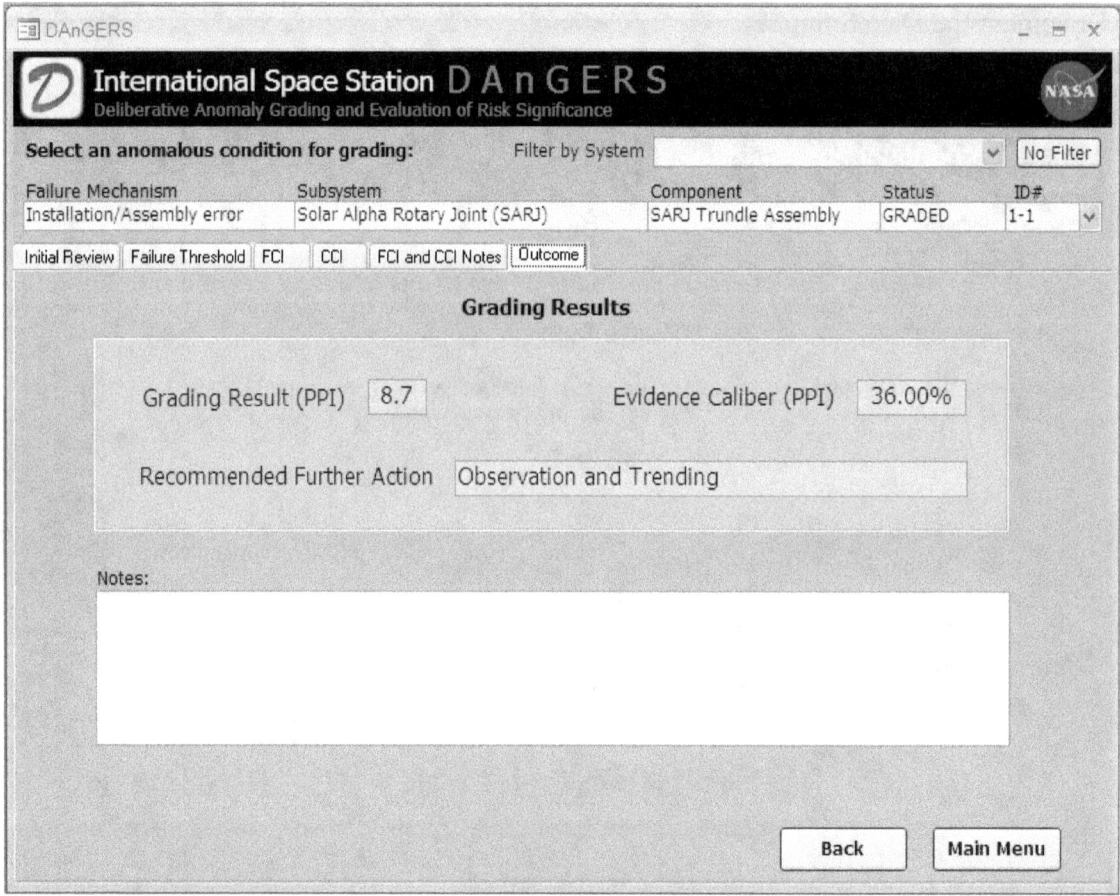

**Figure C-15 Outcome Page**

The recommendation for further action is based directly upon the PPI result. Thresholds set during process development dictate the boundaries between actions; 'Risk Modeling', 'Observation and Trending' and 'None'.

Additional notes on the results of the anomalous condition can be entered into the text box provided at the bottom of the screen, labeled 'Additional Remarks'.

This is the final step of grading the anomalous condition. Navigation to the next anomalous condition in the series can be carried out using the anomalous condition drop-down menu, located at the top of the screen. Whenever a new record is selected, the tool returns to the 'Initial Review' page.

## C.4 Interface with Mission Assurance System

In addition to the DAnGERS features detailed in the previous sections, prototype development has also been completed on a means of integrating the APA process with existing NASA mission assurance processes via common data management systems. The Mission Assurance System (MAS) is a web-based tool originally developed by the

Constellation Program, but has now been adapted by programs and centers including ISS, JSC, and KSC to manage Problem Reporting and Corrective Action (PRACA), Failure Modes and Effects Analysis (FMEA), and other types of Safety and Mission Assurance (S&MA) data. PRACA records are stored within a component of the MAS called the Problem Analysis Reporting Tool (PART). The addition of a precursor capability to this web-based system allows for the integration of precursor information with corresponding anomaly records.

Features established under this effort include data-exchange mechanisms between the MAS and the existing Microsoft Access-based precursor tool (DAnGERS) and a dedicated precursor database within the MAS environment. Additional products of this work include MAS and DAnGERS user-interfaces for the exchange of precursor information, and code development for future adoption into the MAS production environment.

The basic data-flow in the APA process consists of taking anomaly data from a system, analyzing it through precursor evaluation and then producing precursor results. Due to the Generalization step in APA, this produces a 'one-to many' record relationship where one anomaly coming into the process can produce multiple precursor records (called anomalous conditions).

For the MAS and DAnGERS to be considered fully integrated, it requires two separate exchanges of information. One from the PART database within the MAS into DAnGERS; this provides DAnGERS with an anomaly caseload for evaluation. Secondly, data produced during the precursor analysis (anomalous conditions) can then be uploaded to a MAS database dedicated to precursor analysis. The following illustration depicts this data-flow between the three systems; PART, DAnGERS, and Precursor MAS.

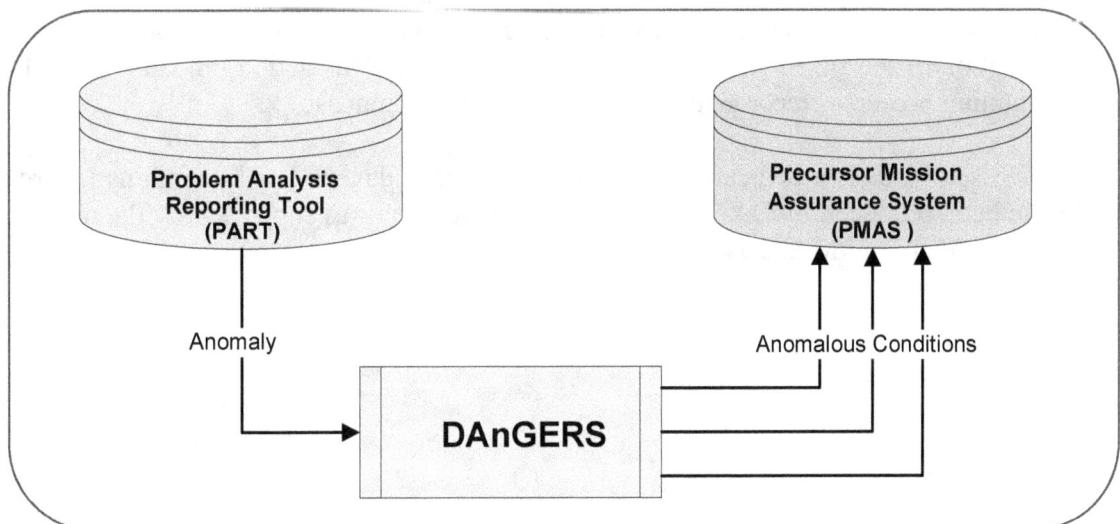

**Figure C-16 Precursor Information Pathway**

### C.4.1 Downloading Records from PART to DAnGERS

The DAnGERS user interface provides a 2-step guide to downloading a caseload of anomalies directly from the PART database and importing those records into the tool. The user must first choose an ISS system and start/end date for the records to be included in the caseload. DAnGERS provides a 'System' drop-down menu and start and end dates fields for selecting these parameters. Once the parameters for the anomaly caseload have been defined, the user can click a 'Download Caseload' button. This begins an automatically download of an Excel file which contains the specified caseload and is formatted for input into DAnGERS. The user may then import this in the standard caseload import method.

### C.4.2 Transferring Precursor Records from DAnGERS to MAS

Once the APA process has been carried out for a set of anomalies, the results can be uploaded to a database within the MAS environment that is dedicated to precursor records, Precursor MAS. Uploading precursor data to MAS is a 2 step process that can be completed directly from the DAnGERS user interface.

Step 1 requires the user to click on an 'Export Data' button in DAnGERS. This causes the tool to export an XML file containing details of the records that have been modified since the last export was initiated. The XML file of precursor records will be exported from the DAnGERS database and saved using a standard Windows 'Save As' dialogue box. Once the XML file is saved, the second step in the process is uploading the XML file to the Precursor version of MAS. Clicking on an 'Upload Data' button in DAnGERS launches the user's browser and resolves a Precursor MAS upload page. Using a 'Browse' button on that page button to launch a File Upload dialogue, the user navigates to the location of the saved XML file and selects 'Open'. This action uploads the XML file containing precursor records into the Precursor MAS database.

The Precursor MAS is a dedicated database to precursor data and only contains records that have been produced by DAnGERS. The layout and features supported by the tool are very similar to current MAS systems, including PART.

# Appendix D - DAnGERS Reports

DAnGERS provides a feature that enables the user to produce standardized reports, enabling the communication of findings to the relevant Risk Manager, to assist in making program decisions relating to safety. The reports module is accessed from the main menu, by clicking on the button labeled 'Reports'

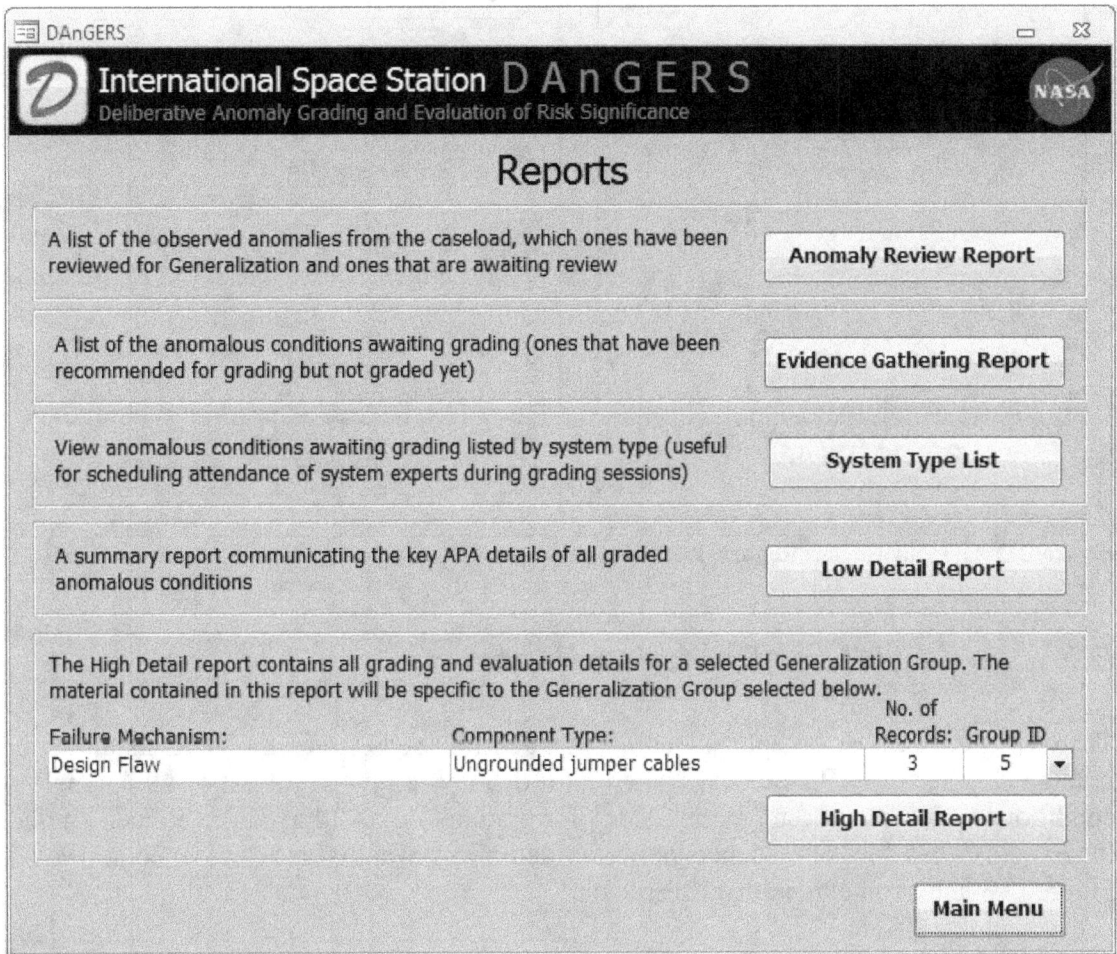

**Figure D-1 DAnGERS Reports Menu**

There are five different types of report that DAnGERS can produce: Anomaly Review Report, Evidence Gathering Report, System Type List, Low Detail Report and High Detail Report.

## D.1 Anomaly Review Report

It is useful to be able to see how many anomalies from the caseload have been reviewed for Generalization, this helps to communicate the level of progress made by a Generalization team at any given time. The Anomaly Review Report lists all the records from the caseload by ID number, Anomaly Title, and PART #; and identifies whether each anomaly has been reviewed or not reviewed.

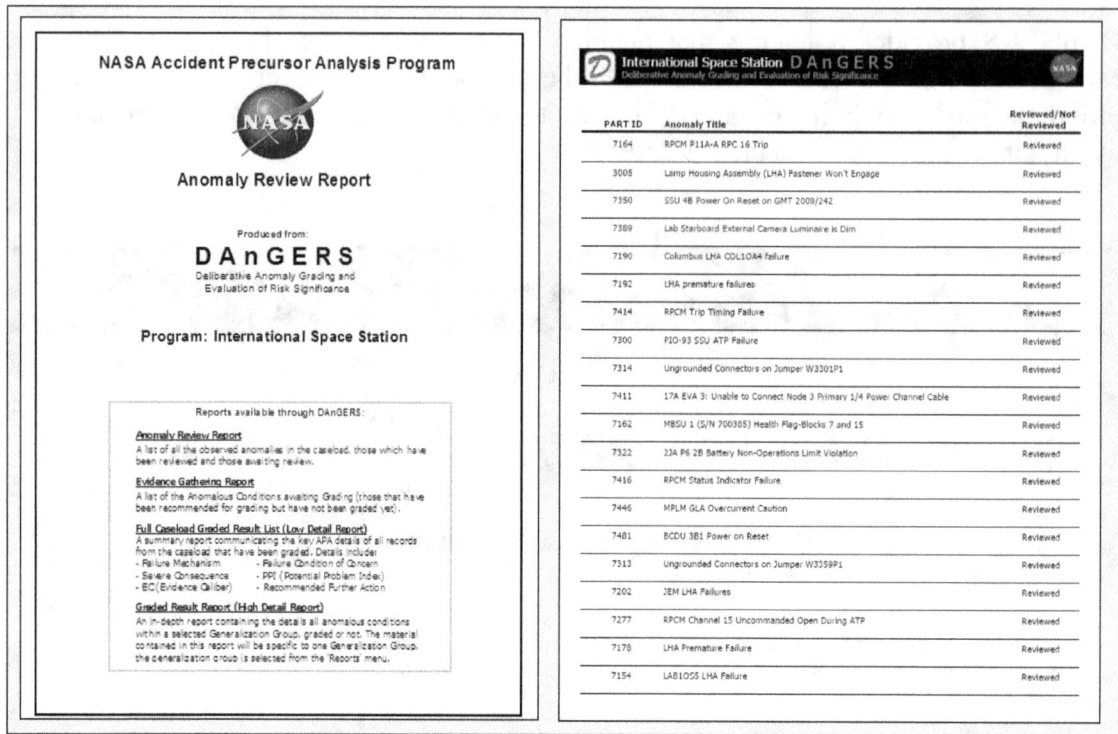

Figure D-2 Anomaly Review Report (ISS Sample)

## D.2 Evidence Gathering Report

The Evidence Gathering Report is report intended for use between the generalization and grading steps of the APA process. The report displays a low-detail list of all anomalous conditions which have been recommended for grading, but that have not yet been graded. It serves as a list for the evidence gatherer, identifying what type of evidence is needed. Figure D-3 shows an example of this report.

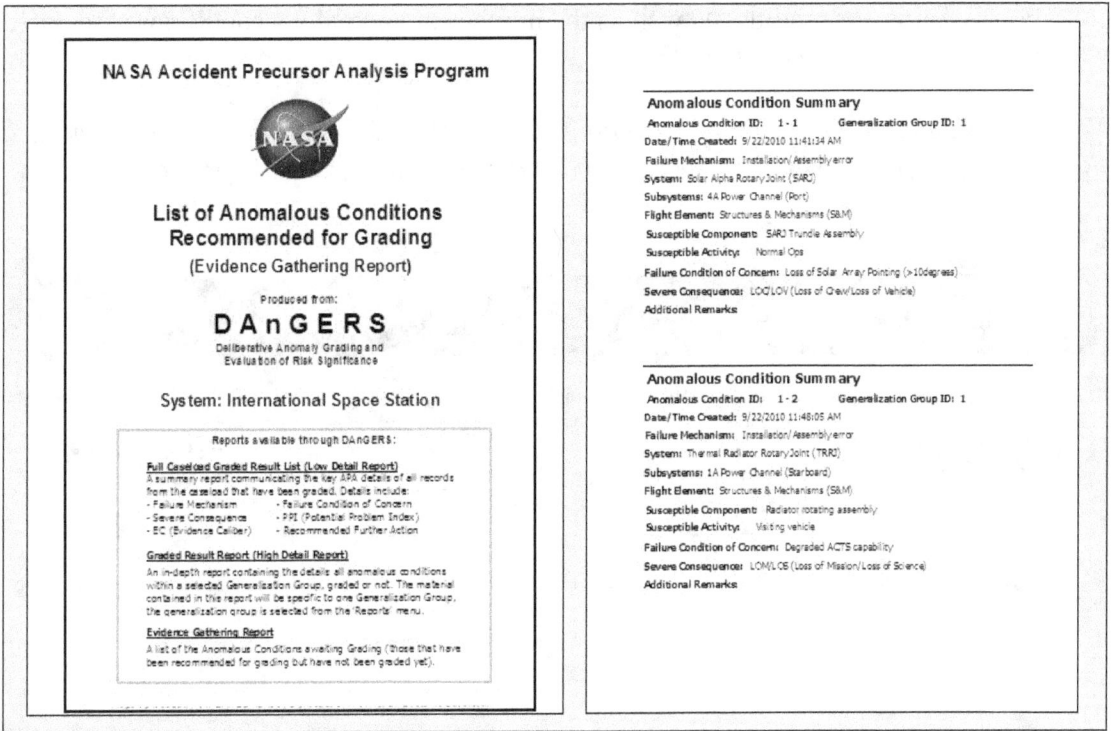

**Figure D-3 Evidence Gathering Report (ISS Sample)**

## *D.3  Scheduling System Experts*

A typical ISS anomaly caseload will contain many records of anomalies related to diverse systems onboard the vehicle. The generalization process expands on this even further to consider the failure mechanism occurring on different components within the scope of generalization. This can result in the generation of anomalous conditions across many systems/subsystems onboard the orbiter. With regards to the grading process, it is prudent to organize the grading sessions to make best use of the resources available. One of those resources to be used effectively is the time of the systems engineers. ISS DAnGERS includes a module designed to aid in scheduling the resources of systems engineers/experts for grading sessions.

Once generalization for a caseload has taken place, the process breaks for evidence gathering, and reconvenes at a later date for the grading exercise. Before grading begins, it is useful for the facilitator or SR&QA manager to know when to include specific experts in the upcoming grading session(s). The 'System Type' module allows users to view a list of all the ISS systems that have anomalous conditions associated with them, in order to schedule the system engineers who are expert in those fields to be present during grading.

The module is launched from the DAnGERS 'Reports' menu, click on the button labeled 'System Type List'. The tool then displays a list of all the systems that have anomalous conditions waiting for grading, as well as the number of anomalous conditions associated with each system type. See Figure D-4.

*Note – The number of anomalous conditions does NOT include those that are not marked 'Recommended for Grading' or ones that have already been graded in previous sessions.*

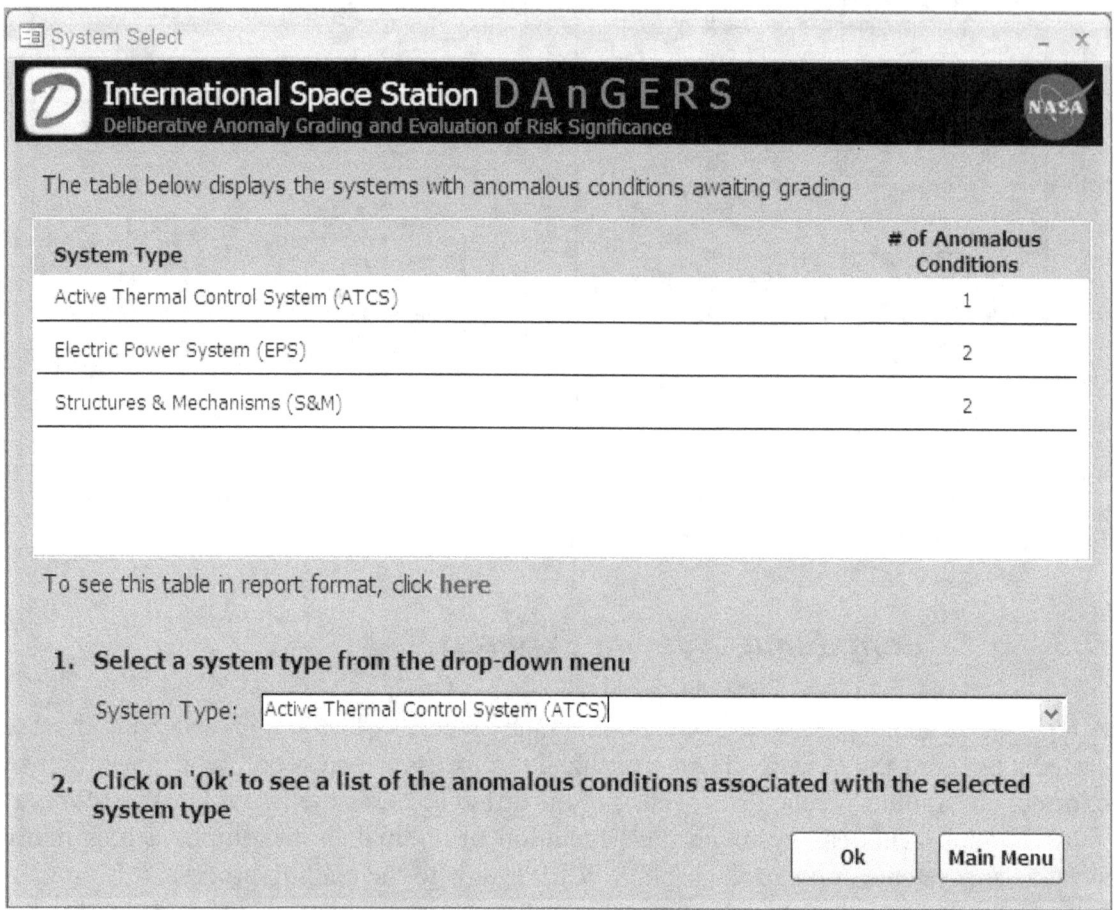

**Figure D-4 System Select Page (Sample Table)**

To see a more detailed break-down of the anomalous conditions associated with a specific system type, follow the on-screen instructions to select a system using the drop-down menu. The detailed list includes information that summarizes each anomalous condition in order to differentiate between each record, the column on the far right (Evidence Collected?) is an optional field for SR&QA personnel to manually record if evidence has been gathered for the anomalous condition.

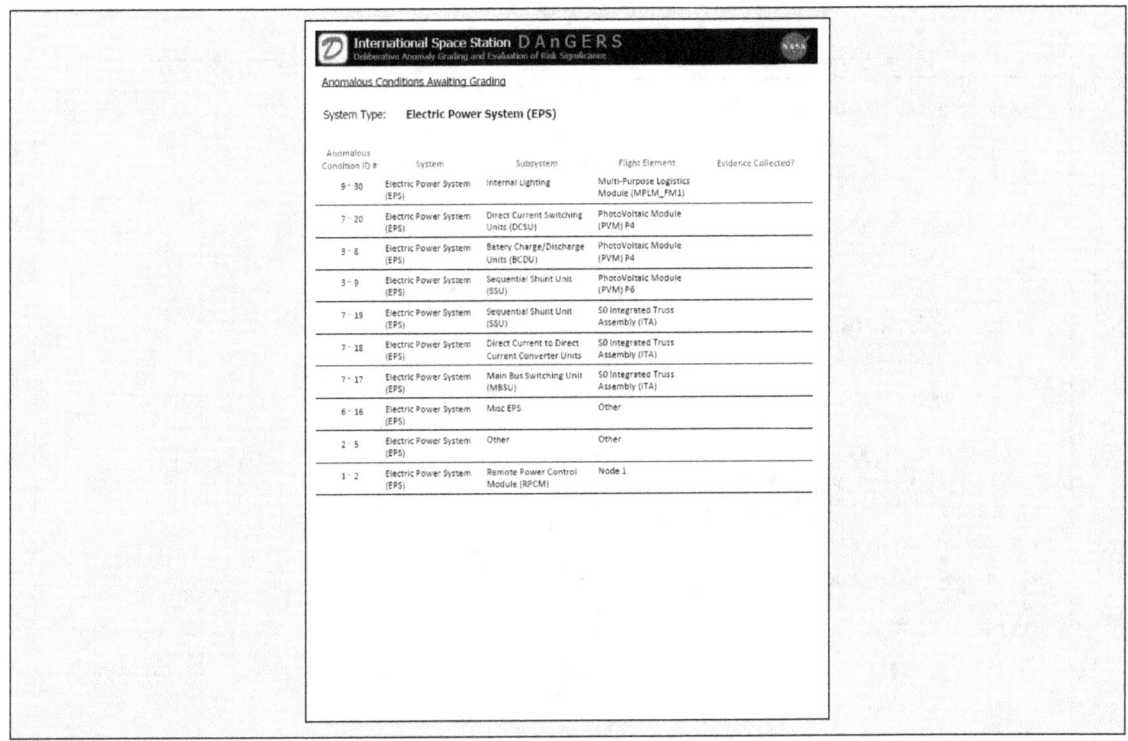

**Figure D-5 Anomalous Conditions Awaiting Grading List (EPS Sample)**

## D.4      *Full Caseload Graded Result List (Low Detail Report)*

The Full-Caseload Graded Result List (Figure D-6) is a low-detail summary report containing the key identifying aspects of each anomalous condition over the full caseload. Details for each anomalous condition include: Failure Mechanism, Failure Condition of Concern, Severe Consequence, PPI, EC, Graded Result (Recommended further action), and several other fields which help to characterize the anomalous condition.

*Note: This report will only display the anomalous conditions that have already been graded. It will NOT include those waiting grading or those NOT marked 'Grading Recommended'.*

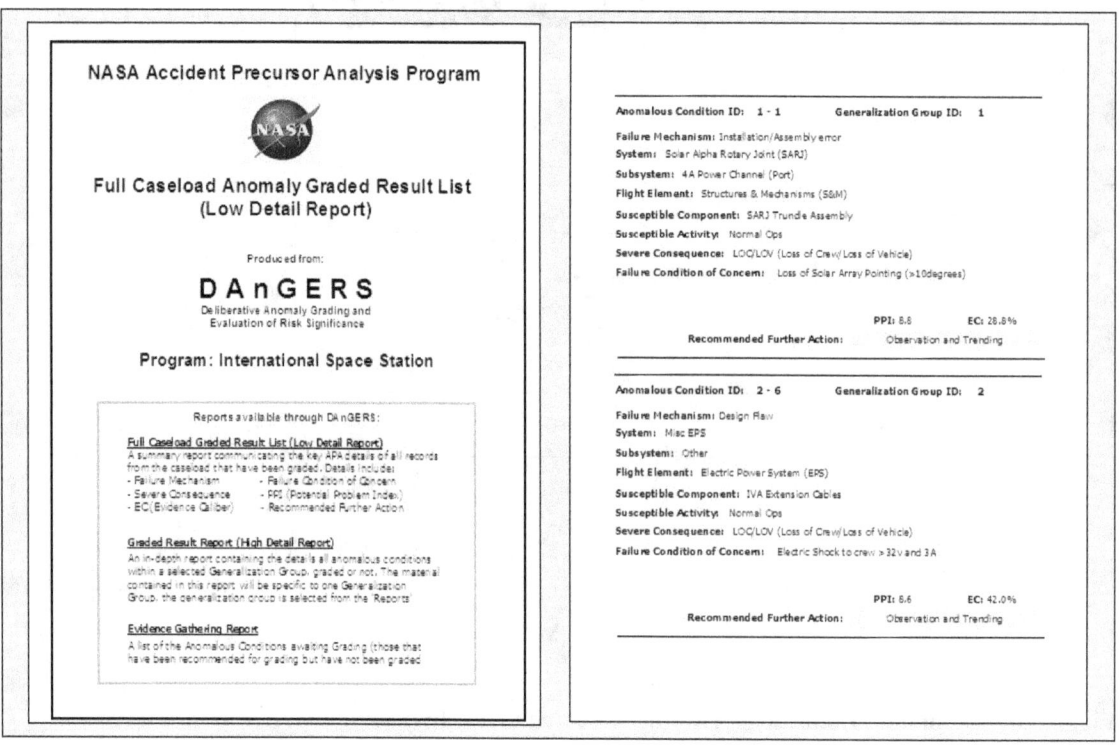

**Figure D-6 Full Caseload Anomaly Graded Result List (ISS Sample)**

## D.5 Graded Result Report (High Detail Report)

The Anomaly Graded Result Report is a high-detail report containing all the details recorded and generated for a generalization group. This report concerns all the anomalous conditions relative to the observed anomalies within the group that served as the kernel for generalization. Users should use the drop-down menu to navigate to the desired generalization group to be shown in the report. The report (Figure D-7) begins with a summary of the observed anomalies within the generalization group, and then lists all the details for all anomalous conditions, including those that were NOT recommended for grading.

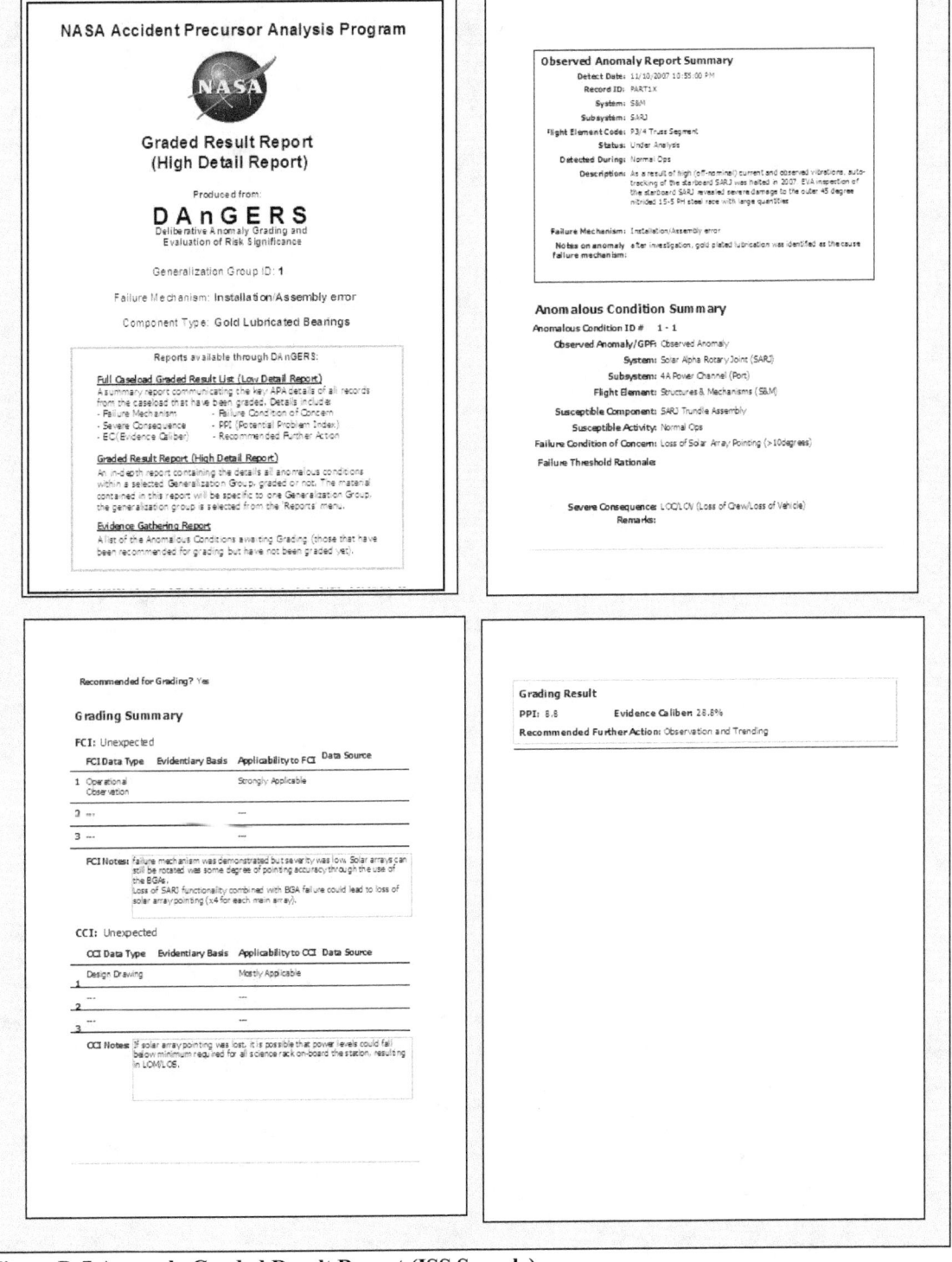

**Figure D-7 Anomaly Graded Result Report (ISS Sample)**

# Appendix E - Defining a Trending Basis

Defining an appropriate trending basis is essential for meaningful trending. In this context, the trending basis can be considered to be the scale by which the data points in a trend are arranged. Simply trending occurrences with respect to a linear time-scale is not always appropriate as we will see in the following examples.

## E.1   Intermittent exposure

NASA systems are typically highly complex involving numerous subsystems, many of which are only in use for limited periods of time when required; robotic manipulators, Extravehicular Mobility Units (EMUs), vehicle docking radar are all examples of subsystems that are only operational during specific periods of time. Subsystems such as these may only experience anomalies when they are operational, and not when dormant. If one were to conduct a trend of such a subsystem's anomaly occurrences with respect to continuous time, the trend shown would not be representative of real-world subsystem usage; the time-scale would include periods of time where the subsystem could not possibly experience an anomaly, which may result in a misleading trend. Within subsystems that are only operational intermittently, it is often useful to trend the anomalies with respect to the period of time in which it is vulnerable to the failure mechanism.

Figure E-1 illustrates an intermittent exposure trend; the top chart shows a series of anomalies with respect to continuous time and the bottom chart shows the subsystem's usage profile (the period of time in which the subsystem is exposed to the failure mechanism). Combining the two into the chart on the right shows how the anomalies are represented along a usage scale, this shows a more meaningful representation of how often anomalies were experienced whilst the subsystem is exposed to the causal mechanism.

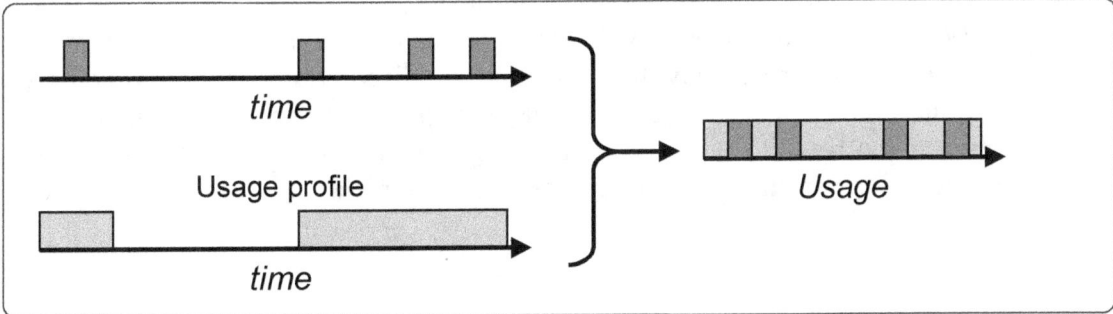

**Figure E-1 Intermittent Exposure Trending**

## E.2 Continuous exposure

Some systems and subsystems are constantly in operation and as such are continuously exposed to potential failure mechanisms that could result in an anomaly. Anomalies in components that are continuously exposed to the anomaly failure mechanism can be trended with respect to their absolute time of occurrence (e.g. ageing, wear-out of continuously-operated hardware).

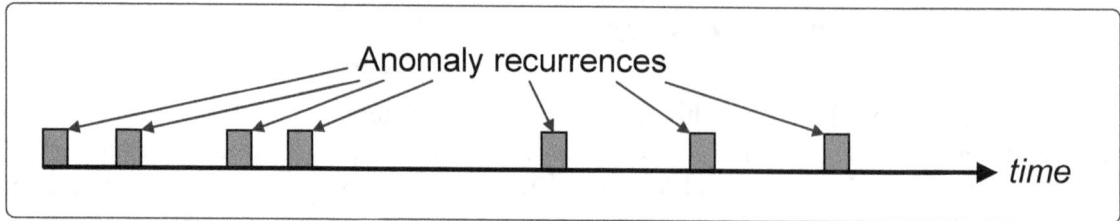

**Figure E-2 Continuous Exposure Trending**

## E.3 Demand exposure

Many components within NASA systems are only functional on an instantaneous basis when they are demanded (e.g. release mechanisms, ignition sources, explosive bolts). Like the intermittently operated subsystems, the anomaly failure mechanism may only be exposed when the component is demanded, and cannot be produced at any other time. Anomaly occurrences within components that are only operational when demanded should be trended with respect to the demands made. In order to appropriately trend anomalies within such components the trending scale may be 'normalized' with respect to the demands on the component.

Figure E-3 demonstrates how a series of demands on a system results in a number of anomaly recurrences, the top chart depicts the anomalies with respect to continuous time and the bottom chart depicts the anomaly recurrences normalized into a demand exposure basis. The resultant trend may reveal previously undetected patterns; anomalies that at first appeared clustered together may become separated and anomalies that at first appeared at regular intervals may then become localized. Some entire systems (in particular, launch vehicles) that are not continuously operated but operated on a mission-by mission basis can benefit from trending analyses that are normalized in this way (e.g. Space Transportation System and other reusable launch vehicles).

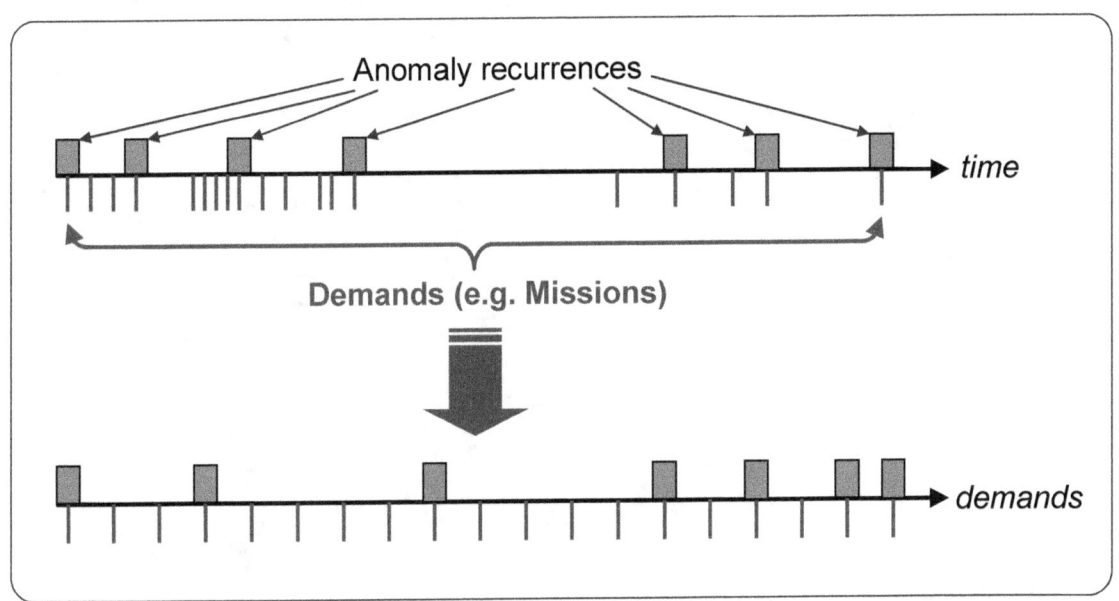

**Figure E-3 Demand Exposure Trending**

# Appendix F - Technical Basis for the Anomalous Condition Risk Index (ACRI)

## F.1 The ACRI Importance Measure

Risk Modeling (in the context of APA) is performed to specifically measure the risk significance of the anomalous condition, and to do so in the context of a benchmark system risk ($R_o$). The APA process defines the Anomaly Condition Risk Importance (ACRI) measure to assist in gauging and prioritizing anomalous conditions in terms of risk. ACRI is the conditional risk that is directly attributable to a failure mechanism occurring outside nominal bounds (thereby creating an anomalous condition). As its name implies, this risk metric is conditional, i.e. it presumes the anomalous condition exists. The ACRI is calculated as:

$$ACRI = [(Risk \mid AnomalousCondition) - (Risk \mid \neg AnomalousCondition)] / R_o$$

**Equation 6**

The second term in the equation is needed to remove any stochastic risk that is represented in the model but not attributable to the anomaly, such as that due to random failures of modeled safety systems. This is done not only to isolate the risk that is directly attributable to the anomaly event, but also to remove the effects of variability in model scope. For example, a detailed model that includes all the components of a subsystem will typically show more risk than a subsystem model that is restricted to the components directly involved in the anomaly.

In the case of ACRI, normalization with respect to the overall system risk allows the significance of anomalous conditions to be assessed relative to other risks in the system, which supports prioritization of risk management attention among competing issues. Particularly, higher ACRI values indicate a greater need for assurance activities in order to maintain baseline risk levels. It also provides a system-independent means for designating an anomaly as a precursor, in cases where precursor criteria are established. For example, an ACRI value of 1% or greater could be considered a reasonable basis for precursor designation.

It is worth noting that ACRI is calculated using risks that are conditional on the occurrence of the anomalous condition. Therefore, when comparing measures across anomalous conditions, the probabilities of occurrence of the conditions are not taken into account. The measures have been defined in this way because anomaly investigation and precursor analysis are intrinsically pre-emptive, i.e. the intent is ideally to find and eliminate vulnerabilities upon their first manifestation. While this isn't necessarily achievable in all cases, it means that the measures will maintain their effectiveness in a data-lean environment, i.e. before many anomalies have occurred.

When a system risk model exists, the system risk $R_o$ can be taken from the model. In cases where a system risk model does not exist, $R_o$ must be obtained by other means in order to provide a normalization factor against which risk significance can be established. One possible basis for normalization, in the absence of a calculated risk, is the system risk requirement, which establishes a de facto acceptable risk. Note that assuming an $R_o$ of 0 (meaning the system has absolutely no risk) leads to an ACRI of infinity – this makes sense since a system with no risk should not be experiencing anomalies in the first place. It is imperative that the same benchmark value of Ro be used in calculating ACRI for all anomalous conditions in a particular system, since this is a comparative measure that is intended for use in prioritizing the expenditure of resources to decrease system risk. Therefore, if the benchmark is updated, it should be applied retroactively to all previous ACRI estimates to ensure a common basis.

## F.2 Analytical Basis of ACRI

If the total probability of failure of a system is denoted P(fail); let the condition where a specific anomalous condition exists be AC; and let the situation where a non-anomalous-condition-related failure-causing condition exists in the system be Q. Note that although Q implies that failure is inevitable, it does not imply that the failure will necessarily be due to Q, since a system with both Q and AC might fail due to one or the other condition. By the law of total probability:

$$P(fail) = P(fail | AC \wedge Q) \times P(AC \wedge Q) + P(fail | AC \wedge \neg Q) \times P(AC \wedge \neg Q) + P(fail | \neg AC \wedge Q) \times P(\neg AC \wedge Q) + P(fail | \neg AC \wedge \neg Q) \times P(\neg AC \wedge \neg Q)$$

**Equation 7**

In principle, the risk attributable to the anomalous condition, conditional on its occurrence, is $P(fail | AC \wedge \neg Q)$. However, whereas it is practical to construct a risk model that allows the results to be conditioned on AC vs. ¬AC, it is not practical to construct a risk model that allows results to be conditioned on Q vs. ¬Q. Thus, the quantities that are amenable to calculation are:

$$P(fail | AC) = P(fail | AC \wedge Q) \times P(Q) + P(fail | AC \wedge \neg Q) \times P(\neg Q)$$

**Equation 8**

and

$$P(fail | \neg AC) = P(fail | \neg AC \wedge Q) \times P(Q) + P(fail | \neg AC \wedge \neg Q) \times P(\neg Q)$$

**Equation 9**

Now, since Q implies failure, and ¬AC ∧ ~Q implies success,

$$P(fail | AC \wedge Q) = P(fail | \neg AC \wedge Q) = 1$$

**Equation 10**

and

$$P(fail \mid \neg AC \wedge \neg Q) = 0$$

**Equation 11**

Substituting Equations 9 and 10 into Equations 7 and 8 yields:

$$P(fail \mid AC) = P(Q) + P(fail \mid AC \wedge \neg Q) \times P(\neg Q)$$

**Equation 12**

and

$$P(fail \mid \neg AC) = P(Q)$$

**Equation 13**

Taking the difference yields:

$$P(fail \mid AC) - P(fail \mid \neg AC) = P(fail \mid AC \wedge \neg Q) \times P(\neg Q)$$

**Equation 14**

Since the left hand side of Equation 13, divided by the benchmark system risk ($R_o$) is the definition of anomalous condition risk index (ACRI), we have:

$$ACRI = \frac{P(fail \mid AC \wedge \neg Q) \times P(\neg Q)}{R_o}$$

**Equation 15**

The method for calculating ACRI accepts an error of $P(\neg Q)$ relative to $P(fail \mid AC \wedge \neg Q)$, which does not make a practical difference as long as $P(\neg Q)$ is close to 1, i.e. as long as $P(Q) \ll 1$. This makes intuitive sense, because subtracting out the fraction of cases with condition Q also subtracts out those cases with condition $Q \wedge AC$, some of which might fail due to AC. Thus the method undercounts failures due to AC in proportion to the fraction of total cases that have condition Q.

# Appendix G - Sample APA Results Reporting

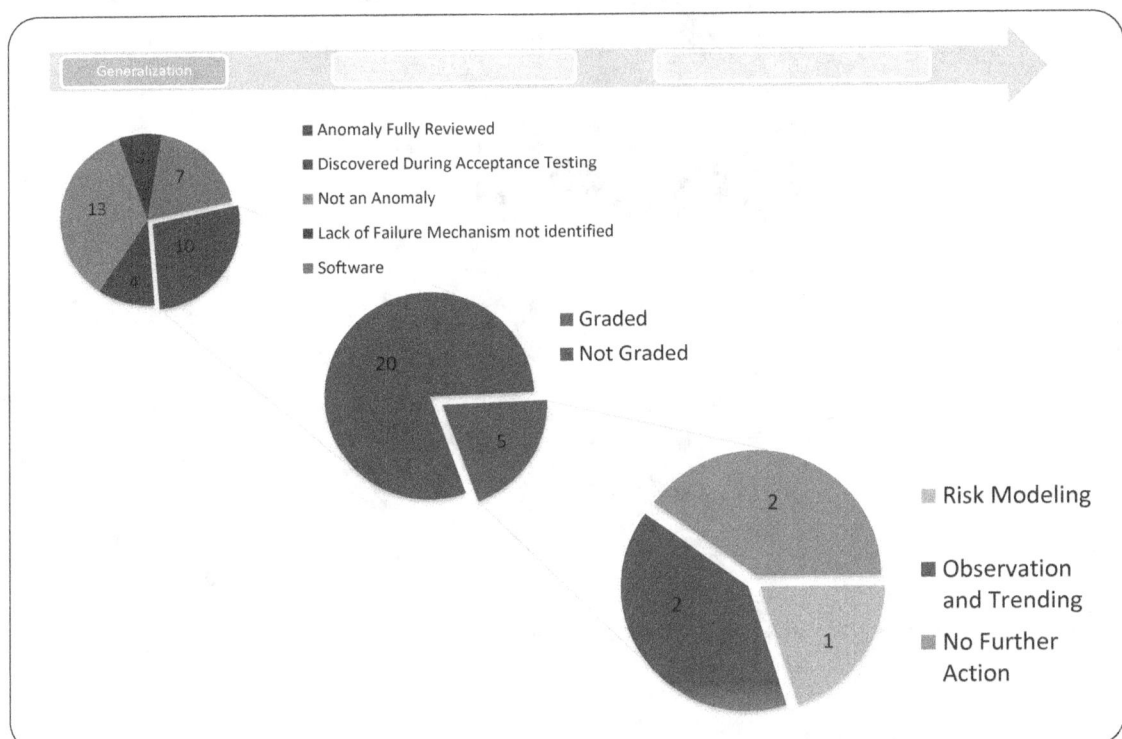

Figure G-1 Sample APA Results

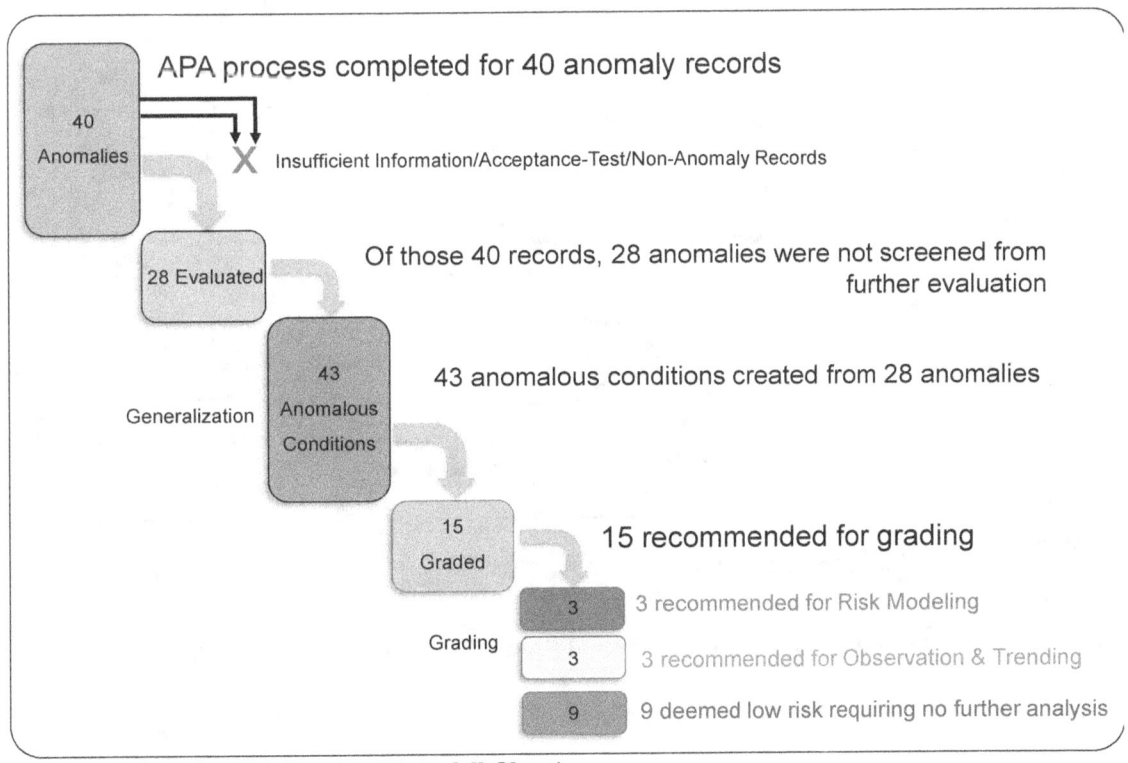

Figure G-2 Sample APA Results (Waterfall Chart)

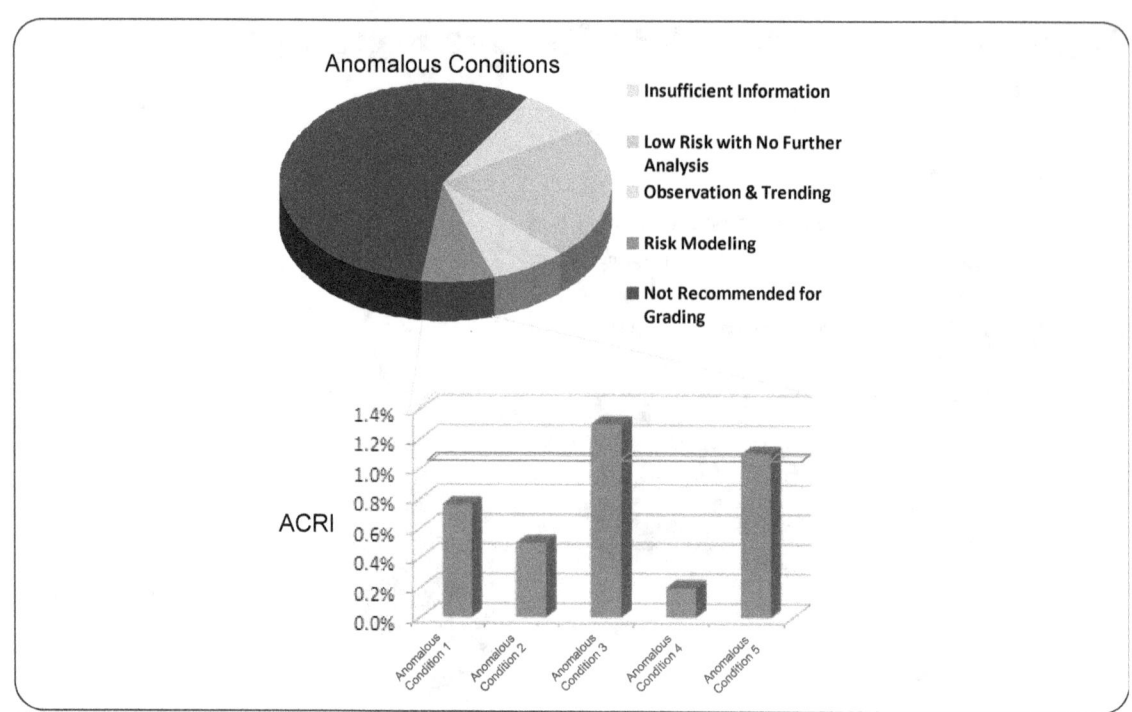

**Figure G-3 Sample APA Results (Risk Modeling ACRI)**

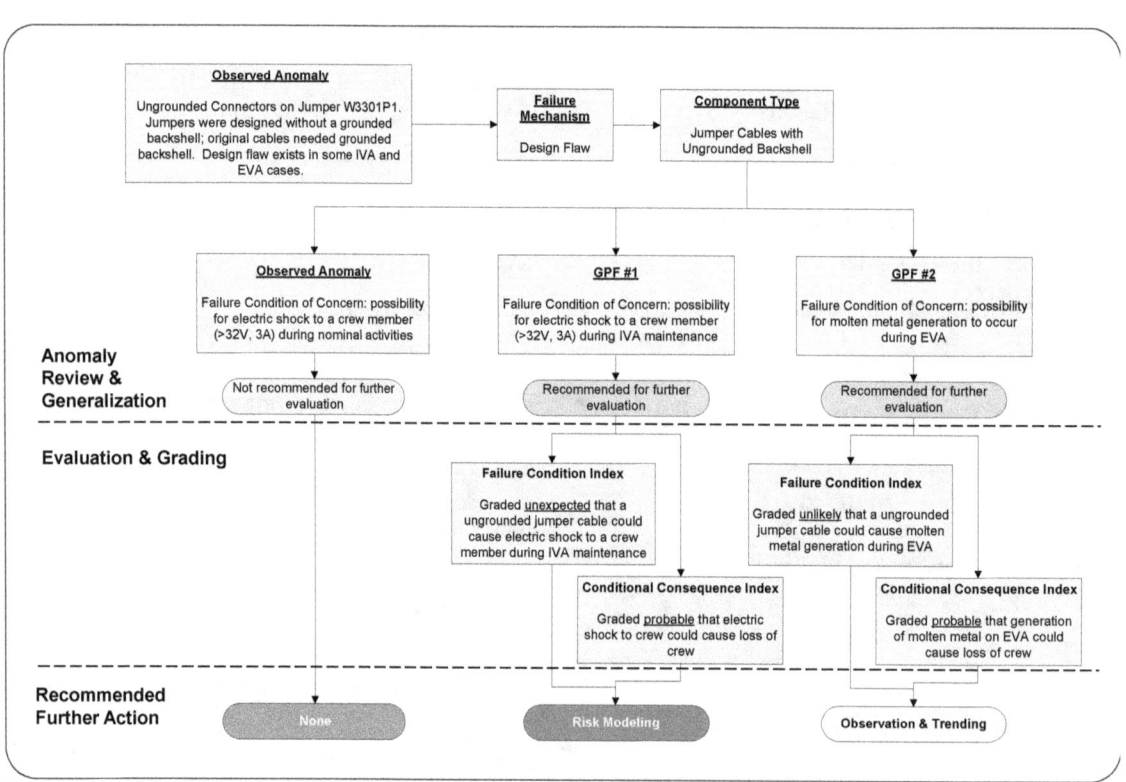

**Figure G-4 Sample APA Results (Process flow and Results of an Observed Anomaly)**

www.ingramcontent.com/pod-product-compliance
Lightning Source LLC
Chambersburg PA
CBHW081729170526
45167CB00009B/3757